A
TREA[TISE]
ON
HEMP.

In Two Parts.

CONTAINING

I. Its History, with the Preparations and Uses made of it by the Antients.

II. The Methods of cultivating, dressing, and manufacturing it, as improved by the Experience of modern Times.

Translated from the French of

M. MARCANDIER,

Magistrate of Bourges.

LONDON,
Printed for T. Becket, and P. A. de Hondt, in the Strand. MDCCLXIV.
[Price One Shilling and Six-pence.]

First published 1996 in facsimile text of 1764 plus illustrations and Part III Postscript by John Hanson/Cht. (Publishers) of the Malthouse, Lyme Regis, Dorset.
© ISBN 0.9529432.0.4

Jos: Banks

Made and printed by Henry Ling of the Dorset Press, Dorchester, Dorset.

TO THE

LAUDABLE

SOCIETY

FOR THE

IMPROVEMENT

OF

ARTS, MANUFACTURES, and COMMERCE,

THIS

TREATISE on HEMP,

As both curious and uſeful,

IS

With the greateſt RESPECT,

Inſcribed by

Their moſt humble Servant,

The EDITOR.

ADVERTISEMENT.

THE Memoir on the preparations of Hemp, which M. DODART, Intendant of Berry, publiſhed in 1755, for the uſe of that province (*a*), excited the curioſity of the publick, with regard not only to the natural properties of that plant, but alſo to its origin and hiſtory.

Though this little work was intended ſolely for the inſtruction of manufacturers, ſome reſpectable perſons, and particularly the Journaliſts of Trevoux, who were pleaſed to give a favourable account of it in the month of January, 1756, ſeemed to wiſh, that, on this occaſion, we had collected all the intereſting particulars, that Hiſtory could furniſh, of the uſe formerly made of Hemp, and the know-

(*a*) Bourges, the capital of Berry, (a province celebrated for its Hemp) was one of the cities of Gaul mentioned by Pliny, where, in his time, ſo great a quantity of linen and Hempen cloth was made, that he even ſeems to reproach them with . . . *Ita ne & Galliæ cenſentur hoc reditu? Monteſque mari oppoſitos eſſe non eſt ſatis, & à latere Oceani obſtare ipſum quod vocant inane Cadurci, Caleti, Ruteni, Bituriges, ultimique hominum exiſtimati morini, immo vero Galliæ univerſæ vela texunt.* Plin. l. xix. c. 1.

ledge

ledge that the nations of remoteſt antiquity had of it. Others, more concerned about the improvement that may be made of it at this time, earneſtly deſired more circumſtantial and particular accounts, not only with reſpect to the culture and methods of preparing this plant, but alſo with regard to the various uſes that may be made of it, in ſeveral ſorts of manufactures. Deſirous to ſatisfy both, we have endeavoured to re-unite, in this Treatiſe, thoſe reſearches and remarks, that ſeem to comprehend almoſt all that Hemp can preſent to the curioſity of the learned, and the utility of the manufacturer: But far from flattering ourſelves that we have exhauſted a ſubject ſo interеſting, and ſo extenſive, with regard to which a great many experiments ſtill remain to be made, we hope, that the firſt effect of this eſſay will be, to excite men of greater ability to take under further conſideration, every thing that has relation to this valuable plant (*a*), and its uſes, which, perhaps, are as yet but very imperfectly known.

(*a*) There is none that offers ſo great advantage to man; it even brings in more than corn. *Nouv. Maiſ. Ruſt.* tom. i. p. 680.

We

We ſhall give ſeveral extracts from different Authors, who have left us any notices of this plant, and its uſes, either in Phyſick, or the Arts; we ſhall alſo give particular obſervations relating to œconomy, which is the principal deſign of this work, and add ſome remarks upon the bad practices, the frauds, the negligences, and abuſes, that have creeped into this branch of commerce. In a word, we ſhall do every thing in our power to attain the end of this work: What further remains muſt be expected from the government, and from the publick.

A TREATISE ON HEMP.

PART I.

Containing its Hiſtory, with the Preparations and Uſes made of it by the Antients.

TO pierce through the dark veil of remoteſt antiquity; to range, with our firſt parents, the fields and foreſts (*a*), and from all the plants, which

(*a*) *Cannabis in ſilvis primum nata eſt.* Plin. l. xx.

When the origin of an art is unknown, we muſt be content with conjecture and hypotheſis inſtead of true hiſtory; and we may be aſſured, that, in ſuch caſes, the romance is more inſtructive than the truth. Generally ſpeaking, chance ſuggeſts the firſt eſſays; they are fruitleſs, and never heard of: Another takes up the affair, he meets with ſome little ſucceſs,

but

cover the face of the earth, to ſelect thoſe which ſeem to have been, at all times, the moſt uſeful and neceſſary; in a word, to diſcover the origin of Hemp, and ſhew how

but ſo little that it makes no noiſe: A third begins where the ſecond left off: A fourth proceeds in the paths of the third; and ſo on, till the laſt of all produces ſome excellent experiments; and only theſe laſt diſcoveries make a ſenſible impreſſion. If we have this diſcovery from a ſtranger, national jealouſy conceals the name of the inventor, and it remains unknown. The origin and progreſs of an art are not like thoſe of a ſcience; learned men communicate their thoughts to one another; they write, they make the moſt of their diſcoveries, they contradict, and are contradicted; theſe diſputes are proofs of facts, and aſcertain their dates. Artiſts, on the contrary, live in obſcurity, and in a kind of ſolitude; they do all with a view to profit, and ſcarce any thing for fame or glory. Some inventions continue whole ages, confined to one family: they are handed down from fathers to their children, and carried to perfection or ſuffered to decay; ſo that we know not preciſely to what perſon, or to what time, we ought to aſcribe their diſcovery. The inſenſible ſteps by which arts riſe to perfection, contribute alſo to render their dates uncertain. *One pulls the hemp, another waters it, a third peels it; it is at firſt a coarſe twine, then a thread, and afterwards woven into cloth; but a whole age intervenes between every ſtep of this progreſs.* Could one man carry a production from its natural ſtate to the higheſt degree of improvement, it would be a very difficult matter to conceal his name. Should a people all at once find themſelves cloathed with a new kind of ſtuff, how could they avoid enquiring to whom they owed the diſcovery? But ſuch caſes never, or very rarely happen. *Encycl. vol.* v. *p.* 647.

mankind,

mankind, from the beginning, have employed it, is an enterprize the more difficult, that historians throw no light at all on this ſubject, and we can by no means find to whom we owe the diſcovery thereof, nor of its uſes.

It is to be preſumed, that this plant having been cultivated, and applied to uſe, long before hiſtories were written, the firſt writers did not think it neceſſary to ſpeak of what was perfectly well known already, and at the ſame time very common.

I ſuppoſe, therefore, that chance or neceſſity, the two great ſprings of invention, at firſt diſcovered to men a plant as common as it is valuable. The firſt who made uſe of it wanted only, perhaps, a withe to tie up a branch, or a girdle for his waiſte. He found, in Hemp, flexibility, limberneſs, and ſtrength. Upon this, he took particular notice of the plant, and obſerved its diſtinguiſhing marks. This was enough to make him communicate it to his family and his neighbours; every one was ſenſible of the utility of this production for all ſorts of ligatures. They would, no doubt, be diſpoſed to multiply and render more common a plant, which appeared ſo very neceſſary:

It was accordingly cultivated; but ſeveral ages, perhaps, elapſed, before any one thought of ſeparating the bark from the ſtem. It was, however, at laſt found out, that by this means its uſe might be rendered more conſiderable and more extenſive. It will be eaſily believed, that the preparation of Hemp, by *watering* it, was not at firſt performed ſo exactly as at preſent. They began, without doubt, with twiſting it into ſmall ropes (*a*), as the ſhepherds in the country ſtill continue to do. Hereafter ſome ſet up rope-yards; others, probably, attempted to ſpin it, then made it into cloth (*b*); but what ſort of cloth muſt this have been? At laſt, as

(*a*) *Cannabis ſativa planta, magni uſus in vitâ, ad robuſtiſſimos funes factitandos.* Dioſcor. *l.* iii. *c.* 141.

Utiliſſima funibus Cannabis. Plin. *l.* xix. *c.* 9.

(*b*) Hiſtory informs us, that the weaver's looms, in ancient times, were of a frame very different from thoſe we uſe at this day. The operators did not ſit, but worked ſtanding; and when they were employed on webs that were to be flowered on both ſides, they turned round their looms.

Arguto tenues percurrens pectine Telas. Æneid. vii.

Homer, Herodotus, Theophylact, &c. acquaint us, that the warp of their hempen cloth was extended from top to bottom of the loom, and that, in applying the woof, which they ſtruck home with a kind of wooden ſword, they worked almoſt in the ſame manner with thoſe that weave girths with us.

by

by degrees, arts, as well as men, are brought to perfection; after ſeveral thouſands of years, Hemp came to be made into fine cloth; and none but perſons of great experience could diſtinguiſh it from what was at that time made of Flax.

Herodotus, the moſt antient hiſtorian, acquaints us, in the fourth book of his hiſtory, "That, in his time, the people of Thrace "cultivated a kind of Hemp, Κάνναβις, that "very much reſembled flax, excepting that the "ſtalk of it roſe higher. Some of it, ſays he, "is cultivated, and another kind grows wild: "Both theſe kinds are preferable to every thing "of the ſort that we have in Greece. The "Thracians (*a*) make cloths of it, which are

(*a*) The Thracians, as we are told by Herodotus, *l.* v. were, next to the Indians, the moſt extenſive of all nations: They derived their original and name from Teras, the ſon of Japheth, who was their Patriarch. In old times under this name were comprehended, not only the inhabitants of Thrace, but alſo the *Getæ*, the *Daci*, and the *Myſians*: The names of *Thrace* and *Scythia* are alſo ſometimes taken indifferently for one another.

Naſcitur autem apud eos (Scythas) Cannabis, Lino ſimillima, præter quam craſſitudine et magnitudine, ſed multo quam noſtra præſtantior, vel ſua ſponte naſcens, vel ſata, ex qua Thraces veſtimenta conficiunt Lineis ſimillima; quæ niſi quis ſit valde

 exer-

" as beautiful to the eye, as thoſe that are
" made of flax; and none can perceive the dif-
" ference, but ſuch as are perfectly acquaint-
" ed with ſuch manufactures.

If it is plain, from this paſſage, that, a long time before the chriſtian æra, Hemp was cultivated in Thrace, as well as in Greece (*a*), and at that time made into fine cloth; may we not conjecture, that other neighbouring nations, with whom they had any correſpondence, underſtood the uſe of it alſo. Can it be ſuppoſed, that the Chaldæans, the Babylonians, the Perſians, and the Egyptians did not make uſe of it, at leaſt for ropes (*b*), which is the firſt uſe to which it would naturally be applied? Is it to be preſumed, that thoſe famous buildings ſo much boaſted of in antiquity, without excepting that ſtately tower, which was the firſt monu-

exercitatus, Linea ſint, an Cannabea, non queat dignoſcere, et qui non viderit Cannabem exiſtimet Lineum eſſe veſtimentum. Herodot. Melp. pag. 281. edit. Græc. Lat. Henrici Stephani, ann. 1592.

(*a*) The Hemp which was cultivated in Greece, was not ſo good as that of Thrace, but it made excellent ropes, as we ſhall ſee hereafter, and, without doubt, coarſe cloth for ſails, and other works of this ſort.

(*b*) *Demiſit ergo eos per funem de feneſtra.* Joſhua chap. ii. ver. 15. This rope is in the Greek tranſlation called σπαρτίον.

monument of the wickedneſs and induſtry of men, were carried to perfection without the help of ropes? Though the holy Scripture mentions only Flax, on all occaſions relating to Hemp, or linen cloth, or garments made of theſe materials, and the Hebrew text has taken no notice of Hemp, under the name we give it, in imitation of the Greeks and Latins, this is no ſufficient reaſon for believing that the Jews were altogether unacquainted with the uſe and properties of this plant. The term λίνον (*a*), or *linum*, which is uſed by

(*a*) The term *linum* and λίνον was uſed to expreſs all ſorts of materials proper for the conſtruction of cloth and ropes, as we are told by Robert Stephens, in his dictionary of the Latin tongue; where he ſays, that λίνον is ἀπὸ τῦ λινέω, *antiquo Verbo, quod eſt teneo, quia Lino omnia tenentur.* On this ſubject he quotes ſeveral authors, who give this word nine or ten different ſignifications*. The word *linteum*, for the ſame reaſon, was uſed to ſignify all thoſe ſorts of cloth, *quæ ex cortice Lini, Cannabis, aut Byſſi, texebantur.* We ſhall ſee hereafter, that the word *Spartum* was of the ſame kind. *Veteribus Græcis* σπάρτον *dicebatur, id omne ex quo fierent vitilia, aut funes, aliaque ad nexum idonea, ut ſunt Linum, Cannabis, Junci, Geniſtæ, &c.* Voſſius, Diction. Etymol.

* Linum pro Filo. *Celſ. l.* vii. *c.* 14.
Pro Fune Nautico. *Ovid.* iii. *Faſt.*
Pro Verriculo. *Virg.* i. *Georg.*
Pro Vinculo. *Id.* v. *Æneid.*
Pro Velo Navis. *Homer. Iliad.*
Pro Linteo in quo dormitur. *Id.*
Pro Hamo Piſcatorum. *Id.*
Pro Fidibus Nervorum. *Id.*
Pro Caſſibus quibus feræ capiuntur. *Ovid.* iii. *Met.*

Hence *Lino Sparton*, not that ſort of which ſheets or ſails were made, but a coarſer ſort of Flax, or Hemp, which was made into ropes.

the Greek and Latin tranſlators; ought to be conſidered as one of thoſe general expreſſions that were much uſed in the Hebrew (*a*) and Chaldaic languages.

The Greeks, indeed, made uſe of a kind of broom, *Spartum*, σπάρτον (*b*), which

(*a*) The Hebrews, for inſtance, uſed the word *Baal*, to ſignify all ſorts of gods or goddeſſes.

(*b*) *In Græcia Sparti copia modo cœpit eſſe ex Hiſpania, neque ea ipſa facultate uſi Liburni, ſed hi plerumque naves loris ſuebant. Græci magis Cannabo & ſtupâ, cæteriſque Sativis rebus à quibus* σπάρτα, *Sparta appellabant.* Aul. Gel. l. xvii. c. 3.

In ſicco præferunt è Cannabe funes. Plin. l. xix. c. 2.

This *Spartum*, σπάρτον, has occaſioned great difficulties among learned men. The Greek authors, and their Commentators, ſpeak ſo differently of it, that it is ſtill very doubtful in what ſenſe it ought to be taken. Some pretend that its name is derived from *ſatum*, id eſt, *ſativum*; while Pliny maintains that the *ſpartum* of Spain *ſponte naſci, nec ſeri poſſe.* Others will have it to come from σπείρειν, *nectere et complicare* (*), becauſe the Greeks gave this name to every thing that could bear to be ſpun and twiſted: This is the opinion of the beſt authors. They uſe the word *Spartum*, to expreſs every thing that is of the nature of Hemp, as the Hebrews did the word Flax to ſignify three ſorts of things, which are often confounded in the Holy Scripture, viz. *Bad, Linum*, the moſt common ſort, which was uſed in the making of cordage and coarſe cloth; *Scheſch, Goſſipium*, a finer ſort, ſometimes taken for cotton, which was uſed in making cloaths for perſons of rank and diſtinction; *Buz, Byſſus*, very fine, which ſerved for making

(*) *Heſychius, Suidas, Ariſtophanes, Pollux, vocant* σπαρτία *vel* σπάρτιον *funes vel funiculos quibus utebantur fabri ad varios uſus, ſive ex Lino, Cannabe, Junco, vel alia materia nexi fuerint, aut confecti.* Henr. Steph.

they brought from Spain, for the ſervice of their marine, and for caulking their ſhips, becauſe it made greater reſiſtance to the water than Hemp; but for ropes and every other purpoſe they preferred Hemp. Is it poſſible that Nineveh, Babylon, Memphis, Palmira, Thebes, and ſo many other famous cities ſhould have been quite unacquainted with the uſe of a plant ſo neceſſary and ſo common?

The Romans made ſails of it, and ropes for their ſea (*a*) and land ſervice. They had magazines of it in two of the principal cities of the weſtern Empire; the Hemp neceſſary for the purpoſes of war, was, by the Emperor's orders, amaſſed at Ravenna in Italy, and

making ornaments for the uſe of the Prieſts and the Temple; it is ſcarce poſſible to imagine, that the Jews were entirely unacquainted with Hemp, which was ſo well known to other nations (*). It is much more natural to ſuppoſe, that it was one of theſe kinds of Flax we meet with in the ſeveral texts of Scripture, where mention is made of coarſe, and fine linen, and of cordage. See Ezek. ch. xxvii. ver. 16; 1 Chron. ch. vi. ver. 21; ch. xv. ver. 27; and 2 Chron. ch. ii. ver. 14; ch. iii. ver. 14; Eſth. ch. i. ver. 6; ch. viii. ver. 15.

(*a*) *Tun' mare tranſilias tibi torta Cannabe fulto. Cœna ſit in tranſtro.* Perſ. Sat. v. ver. 146.

(*) *Eſt enim vero Eleorum ager, & cætera ferax, & Byſſum educat feliciſſime. Cannabem quidem Linum & Byſſum ſerunt, qui idoneum ad Lac ferenda ſolum colu.t.* Pauſan. lib. vi. ad finem.

Vienna

Vienna in Gaul. The officer, who ſuperintended that matter on the further ſide of the Alps, was called the Procurator of the Hemp manufactures in Gaul, and had his reſidence at Vienna. Their huſbandmen uſed Hemp for tying the oxen to the yokes (*a*); and, without doubt, for all the other purpoſes relating to agriculture. We know that they made no great uſe of linen cloth, but they had ſome; and Vigenerus, upon Titus Livius, tells us, that they made uſe of Hemp: Even their laws and their annals were written on hempen cloth (*b*). Nothing is more generally or better known than the uſe they made of it for adorning their theatres, covering their ſtreets and their publick places, their amphitheatres, and their *arenæ* for the Gladiators, to ſhade

(*a*) *Cannabiniſque funibus cornua jumentorum ligato.* Columel. lib. xvi. cap. 2.

(*b*) *Licinius Macer auctor eſt & in fœdere Ardeatino, & in Linteis Libris ad Monetæ inventa... quæ ſi in ea re ſit error, quod tam veteres Annales, quodque Magiſtratuum Libri quos Linteos in æde repoſitos Monetæ, Macer Licinius, citat identidem Auctores.* Tit. Liv. l. iv. c. 7. & 20.

It was in the temple of Moneta that the books of Hempen cloth, containing the deſtiny and fates of the Roman Empire, were kept with great care.

The Samnites alſo made uſe of the ſame cloth for the purpoſe of writing. Tit. Liv. l. x.

those

thoſe who aſſiſted at their publick ſhews. Plin. l. xix. c. 1.

Martial informs us, that the Romans alſo made uſe of hempen cloth for table linen, and that every gueſt generally brought his napkin with him (*a*). We cannot therefore in the leaſt doubt, but Hemp was known to the ancients (*b*), as a material of cloth for the ſervice of war, both by ſea and land (*c*), as well as for the purpoſes of agriculture. And if the greateſt part of authors have ſometimes made uſe of the word *Spartum* (*d*), to ſignify ropes; even when thoſe ropes were made of Hemp, the reaſon was, becauſe they conſidered *Spartum* as a general term applicable to Hemp as well as Flax, or any material of that kind (*e*);

(*a*) *Attulerat nemo mappam dum furta timentur.* Mart. l. xii.

(*b*) We ſhall afterwards ſee to how many uſes Hemp was formerly applied.

(*c*) *Ubi vis magna Sparti fuit ad rem nauticam congeſta ab Aſdrubale.* Tit. Liv.

(*d*) *Sparteus generaliter pro quovis funiculo ponitur, ſive è Sparto nexus ſit, ſive è Cannabe, Lino vel aliunde.* Athen. l. v.

(*e*) *Græci Juncos quippe ipſos, & Geniſtas, & quidquid denique ad funes nectendos, & aliquid ligandum verti poſſet, σπάρτον vocavêre: hi autem vocem hanc σπάρτον, de herbis omnibus ad vitilia, nexilia, textiliaque aptis uſurparunt. . .* Salmaſ. Exercit. Plin. pag. 261. . . . and he adds: . . . *ex Lino Hiſpanico, quis putet rudentes navium tortos unquam fuiſſe? Nugatur itaque Solinus,*

nec

unleſs the ſignification of it be otherwiſe abſolutely determined. In a word, how much ſhould we know of the uſe that was formerly made of Hemp in China and Japan, and ſo in both hemiſpheres, if their hiſtories had come with more accuracy into our hands.

We read in Kolben, that the Hottentots uſe a plant, named *Dakka*, inſtead of tobacco, or at leaſt mix them together, when their proviſion of the latter is almoſt exhauſted. This herb, ſays he, is a kind of wild Hemp, which the Europeans ſow principally for the uſe of the Hottentots (*a*).

nec enim ad id dixit Mela. Ex Lino tamen armamenta navium etiam olim fuiſſe, eruditioribus placuiſſe, ibidem notat Plinius, qui verſum Homeri ita interpretabantur, quoniam cum ſparta dixit ſignificaverit ſata. Quæ non intelligo, quaſi neceſſe ſit σπάρτον, nomine Linum accipi, quia ſignificaverit ſata. An non & Cannabis ſativa, de qua τὰ σπάρτα, id eſt ſata, in illo Homeri loco poſſumus interpretari. . . . Nugatur itaque Solinus, niſi dicamus eum ſub materie rudentum, Spartum tantum comprehendiſſe.

The *Spartum* of Spain, which we interpret *broom*, is a ſort of ruſh, *Juncum aridi ſoli*, that grows near Carthagena; it is prepared almoſt in the ſame manner with Hemp by *watering*; it naturally grows in that ſoil, and cannot be raiſed from ſeed.

The Grecks alſo made uſe of another kind of ruſh for making ropes, which from thence had the name of χοῖνος.

(*a*) *Hiſtoire Generale des Voyages*, l. xv. *Hiſt. Nat. du Cap, tirée de Kolben.*

Though

Though the etymology of the word Hemp does not appear to be a matter of great importance, we think it our duty not to omit it, that we may leave nothing untouched that can be propoſed with regard to this plant. Some pretend that it comes from the Celtic word *Canab* (*a*); others think they find its root in the Greek word Κάννα, or Κάννη (*b*), which comes from the Hebrew *Kannab*; in Latin *Canna*; in French *Canne*; becauſe its form, and the length and ſize of its ſtalk give occaſion to compare it to a cane. And therefore, as we ſay, a ſugar-cane, &c. we may alſo ſay, a hemp-cane. The terminations of particular languages, ſuch as Κάνναβις or Κάνναβος, in Greek; *Cannabis* or *Cannabum* in Latin; *Canapo* in Italian; *Canamo* in Spaniſh, &c. are ſo many expreſſions peculiar to thoſe languages, but that variety makes no alteration in the ſignification.

Hemp is commonly diſtinguiſhed into two ſorts; one that grows wild, *Cannabis ſilveſtris*; the other produced by cultivation, *Cannabis*

(*a*) Pezron. *Cannabis, Græcè* Κάνναβις *vel* Κάνναβος, *unde & Belgicum* Kennep, *quaſi* Kannab, *herba eſt funibus faciendis idonea, à Lino, & tenuitate, & candore diſtans. Eſt verò* Κάνναβις, Κάννα.
(*b*) Voſſius.

domeſtica.

domeſtica. This latter is of two ſexes; the male *fructifera*, and the female *florifera*; but both improperly ſo called, for it is more natural to call that the *female*, which bears the fruit, than the other that bears the flower. The ſeed and the root of wild Hemp, are like thoſe of the wild mallow; the ſtalks are ſmaller, blacker, more rough, and about a foot and half in height; the leaves are like thoſe of Hemp raiſed by cultivation, but more rough, and likewiſe blacker (*a*).

The root of Hemp produced by cultivation is ſix inches long, or thereabout, of a whitiſh colour, ligneous, undivided, and running to a point, having fibres only on two lines, diametrically oppoſite to one another, when it is not ſtraitened for want of room; and thick in proportion to the ſtalk it bears. The ſtalk is round, from the root to the firſt ramification; it then aſſumes a quadrangular form, and is fluted, hollow, ligneous, covered with a greeniſh bark, compoſed of filaments, hairy and rough to the touch: at proper diſtances this bark is ſecured from place to place by ſix ſmall faſtenings, which keep it cloſe to the

(*a*) *Nigriore folio & aſperiore.* Plin. l. xx. c. 23.

ſtem,

ſtem, like ſo many little nails, regularly ranged on the circumference of the ſame circle, and almoſt equally diſtant from one another. Its length and thickneſs are various, according to the difference of the ſoil, of the method of cultivation, of the climate, and of the ſeaſons. Some of it riſes to the heighth of eight or ten feet, and the ſtalks look like ſo many little trees (*a*); others ſeem to pine away on the ground, and ſcarce get to the height of two or three feet, ſometimes leſs. A grain of Hemp-ſeed, ſown by itſelf, in a ſoil that agrees with it, commonly produces a ſtalk very large (*b*) and firm, with many branches, and looks like a little tree. If it is of that ſex which produces ſeed, it will yield a great many grains, and thoſe very beautiful: But its bark being too hard and thick, will not be very fit for manufactur-ing: On the contrary, ſeed ſown in a field, that is properly prepared for the purpoſe, and near to each other, produce ſtalks that are ſtreight, ſmooth, without branches, ſofter (*c*),

(*a*) *Quod ad proceritatem attinet, Roſea agri Sabini arborum altitudinem æquat.* Plin. l. xix. c. 9.

(*b*) Of ſuch ſtalks they make a kind of charcoal, fit to enter into the compoſition of gunpowder.

(*c*) *Quo denſior eo tenuior.* Plin. l. xix. c. 9.

and

and more tender than the former; and the bark of ſuch being ſmooth, fine, and ſoft, is much valued for ſeveral uſes. The leaves grow two and two oppoſite to one another; they are divided into many ſegments, being narrow, oblong, ſharp pointed, jagged, full of veins, of a deep green colour, rough to the touch, and of a ſtrong ſmell, that affects the head.

The flowers that grow on the female ſtalk, as it is commonly called, iſſue from the *alæ* of the leaves, on a pedicle of four little cluſters, lying in the form of a Saint Andrew's croſs: They have no *petals*, and conſiſt of five ſtamina, with yellowiſh ſummits, in a calyx of five leaves, of a purple colour without, and whitiſh on the inſide; theſe flowers are not followed by any fruit; and, on the other hand, the fruit on the ſtalks that produce it, is never preceded by any flower.

Whatever the order of nature may be in the vegetation of this plant, both the male and the female ſtalks are produced indiſcriminately from the grains of ſeed that grow on the ſame ſtalk; and the difference cannot be known till they come to bloſſom. We know not, when we ſow Hemp, what quantity of either ſex will

will be produced, nor which contributes moſt to the propagation of the plant; they cannot, however, be eaſily diſtinguiſhed till (*a*) ſixty days after they are ſown; but this obſervation, hitherto, does not appear to have been of any conſequence.

The fruit grows in a great number of bunches at the end of the ſtalks and branches, which naturally produce them: This fruit is terminated by a forked ſtyle, when it is in *embryo*, and is wrapt in a membrane, which ſecures it till it comes to maturity; then the piſtillum changing to a roundiſh grain, forces the membranous capſulæ, which contain it, to open, and we therein diſcover a round ſmooth grain, ſomewhat flattened, and of a ſhining grey colour, containing, under a thin ſhell,

(*a*) The time when the Hemp bloſſoms cannot be exactly aſcertained; for it depends on ſeveral circumſtances. Sometimes it is not above a foot high; when this is the caſe, the Hemp continues weak, and grows little or nothing at all higher; this is ſometimes occaſioned by great heats, or other unfavourable accidents; at other times it riſes to the heighth of four or five feet before it bloſſoms, and grows almoſt as much after. The Hemp, which bears the flower, commonly gets before that which produces the ſeed, and riſ s about half a foot above that which bears the grain. This ſuperiority, in the order of nature, may be well accounted for, if it is true, that the powder, which iſſues from the flower, ſerves to convey fertility to the grain on the ſtalks that bear the ſeed.

a tender, ſweet, and oily, white kernel, of a ſtrong ſmell, that intoxicates when it is freſh. This kernel is covered with a green pellicle, terminating in a point on the ſide next the germ, which is very ſingularly ſituated.

This grain, which is called *Hemp-ſeed*, is no leſs uſeful for its peculiar qualities, than for thoſe which it has in common with the whole plant. Its ſubſtance, conſidered as a ſeed, is ſoft, fat, oily, and gummy; it ferments, conceives heat, and ſprings up with equal facility; its pores being large, tender, and flexible, receive greedily the impreſſions of heat and humidity, which tranſmit to them the nutricious juices ſupplied by a fat, light, and well-laboured ſoil; its fibres, after a quick germination, unfold themſelves, grow up, and attain ſtrength, and the gum, being the principle of their union, ſupports and preſerves them. Beſides the uſe of its oil in phyſick, it is alſo employed, with great advantage, in the lamp, and in coarſe painting: They give a paſte made of it to hogs and horſes, to fatten them; it enters into the compoſition of black ſoap, the uſe of which is very common in the manufactures of ſtuffs and felts; and it is alſo uſed for tanning nets.

A grain

A grain of Hemp-ſeed, ſeen by the help of a microſcope, preſents at firſt a greyiſh epidermis, full of veins, the compartments whereof appear like a ſort of ſcales. Under this firſt cover you ſee a brown olive coloured bark, extremely ſmooth on the inſide, formed of two ſhells, which ſeparate exactly in the middle, like thoſe of a nut, the ſeam that joins them being quite imperceptible. Under a green cover, its kernel, in the form of a little orange, bears its germ produced along one of its ſides, which makes it look a little flatted: When you have taken up this pellicle, you find a white kind of matter, conſiſting of two lobes joined together, which evidently form a kind of head; theſe lobes are very diſtinct, and by the germination are made to ſwell, open, and ſeparate. Its germ, which is roundiſh, bending back along the whole external length of the grain, under the ſeam which joins the two ſhells, terminates in a point, and forms a kind of tendril, which is the only part that pierces the ground to form the root; the other end of the germ, which lies concealed between the two lobes that encloſe and preſerve it, appears like an exceeding fine and delicate ſort of lance (*a*); from it iſſue the two leaves that

(*a*) Which they call *the feather.*

appear firſt, and we may imagine it to be the true principle of its germination and life. Theſe two lobes are alſo changed into two ſorts of thick green leaves (*a*), of an oval form, but not indented, which ſerve for a rampart, and preſervative to the ſpringing leaves. The whole of this white matter ſeems to be fat and ſpongious; and its pores appear to be no leſs open than thoſe of ſnow: And it is, no doubt, owing to the ſituation of its germ, and the ſoftneſs of its whole ſubſtance, that Hemp-ſeed, beyond any other ſort of grain, has ſo great a diſpoſition to ferment, and ſpring up almoſt as ſoon as it is ſown.

The bark, as it appears upon the ſtalk, forms a green, knotty, rough or prickly covering to it. Theſe knots and prickles are mere excreſcences of gum, of which the whole bark is compoſed; but they have different degrees of force and adheſion. This firſt ſuperficial gum ſerves only to keep the fibres of the Hemp cloſe together, and as a kind of maſtick to cover, ſtrengthen, and protect them, againſt the inclemency of the air, the duſt, and the rain: It diſſolves, exfoliates, and breaks, when the bark is watered.

(*a*) Which the Botaniſts call *the ſeminal leaves.*

The

The inſide of the bark, which touches the ſtem, is ſmooth, ſoft, and white; the fibres are very diſtinct from one another, and appear perfectly in all their dimenſions, by means of the watering juſt mentioned. It was not obſerved in former times, that the thread had its exiſtence in the plant, without any dependence on the operations of art; that the labour is confined to cleaning, dividing, and ſeparating the ſoft fibres of which the bark is compoſed; and that this bark is a kind of natural ribbon, or ſcarf, the threads whereof are applied and joined together, lengthways, only, by a dirty glutinous humour, which muſt abſolutely be diſſolved and ſeparated, becauſe it is equally hurtful to the workmen and the work. The threads themſelves alſo conſiſt merely of a gum, but of one that is of a different quality from the ſuperficial gum; they are ſupple, ſtrong, and reſiſt the impreſſions to which the former give way. Every fibre is compoſed of gummy globules, that are very fine, tranſparent, and bright, when ſufficiently cleared from that ſuperficial gum that ſurrounds them; and which the microſcope ſhews to be of a different ſort. All this will appear plain, if you take a few fibres from a thread that is thoroughly bleached. The

 fibres

fibres of Hemp, in this ſtate, are nothing different from thoſe of cotton and ſilk, which makes it reaſonable to conſider them as materials of the ſame kind: And it is a convincing proof of this, that when they are mixed and carded together, there appears to be a complete ſameneſs in the whole mixture.

We ſhould have found, without doubt, more curious and circumſtantial obſervations, in the generality of authors who have examined this plant, if they had been as fully perſuaded of its utility in the arts, as of its medical properties.

Pliny tells us, that Hemp-ſeed is of a drying nature, that it weakens the generative powers in men *(a)*, when they eat it to exceſs. On the contrary, it promotes fruitfulneſs in fowls, for which reaſon it is purpoſely given them in winter time, and is a food to which birds are accuſtomed. It expels wind; is hard of digeſtion; and diſagrees with the ſtomach; it produces bad humours, and occaſions headachs *(b)*. It was formerly one of thoſe legumes, which were fried for deſerts *(c)*: It

(*a*) *Semen ejus extinguere genituram virorum dicitur*. . . . l. xx. c. 23.

(*b*) *Sed cum dolore capitis*. Ibid.

(*c*) De la Mare, *Traité de Police*.

was

was alſo made into little ſweet cakes, to be eaten at collations, and to promote drinking; but, at preſent, this unwholeſome ragout is quite baniſhed from our tables: It heats thoſe that eat it too freely ſo much, that it occaſions very dangerous vapours *(a)*; ſo that thoſe who preſcribe a decoction of this ſeed to children that labour under epilepſies, far from procuring them relief, increaſe and irritate their diſorder. The juice of it *(b)*, ſqueezed out when it is green, draws inſects to it, and brings out all the vermin that enter into the ears, and infeſt them. Taken in an emulſion *(c)*, it is good againſt a cough and the jaundice, and alſo againſt the gonorrhœa; its oil is recommended as an ingredient in pomatums for the ſmall-pox; and it is laxative. Taken inwardly, or outwardly applied, it has not the dangerous qualities that are aſcribed to the whole plant with its leaves; the powder of it, mixt with drink, will make thoſe who uſe it drunk, dull, and ſtupid: We are told that

(*a*) Gallien, lib. vii. *de Simpl. Medic.*

(*b*) *Succus ex eo vermiculos aurium, & quodcumque intraverit, ejicit.* Plin. l. xx. c. 23.

(*c*) We meet with this in ſeveral authors who boaſt of its effects. See *Emulſio Cannabina ad Gonorrhœam*, de Doleus, Etmuller, Michaëlis, & Minſchit, &c.

the Arabians *(a)* make a ſort of wine of it, which intoxicates, and poor people eat the oil of it in their ſoup.

The grain and the leaves being ſqueezed, while they are green, and applied, by way of cataplaſm, to painful tumours, are reckoned to have a great power of relaxing and ſtupifying. The ſmell of it is extremely ſtrong and intoxicating. It is pretended, that the water, in which Hemp hath been ſteeped, proves a deadly poiſon to any that drink of it: This may be true; but what is commonly ſaid of the ſame danger to fiſhes in the rivers *(b)* and ponds, in which Hemp is watered, is falſe. Fiſhes love this plant, and fly to it; but if any accidents of this kind ſhould happen, they can only be owing to the ſmallneſs of ſome reſervoirs, where the water not having a free courſe, may be too much impregnated with the juice of the Hemp, or afford to the fiſh too much of a delicious food, an exceſs of which is always hurtful.

(*a*) De la Mare, *Trait. de Pol.* l. v. tit. 15. Where he quotes Simeon Sethi, *De aliment. facul. C. Apitii, de re culinar.*

(*b*) The Ordonance of Waters and Foreſts on this ſubject ſeems not to be well founded.

What

What Pliny aſſures us, of the great effect which an infuſion of Hemp may have in coagulating (*a*) water, will not appear ſurprizing if we attend to the quality and quantity of the gum, which unites all the fibres of this plant together, and whereof, in reality, it intirely conſiſts. It is, doubtleſs, for this reaſon, that it is given in drink to cattle to cure looſeneſs. The decoction of green Hemp, with its ſeed, when well cleared of the dregs, cauſes the worms to come out of the ground on which it is poured, and the fiſhermen commonly make uſe of this expedient to catch them, when they have occaſion.

Matthiolus preſumes, that it may alſo have power to drive worms out of the human body. It is given in a drink to cattle and horſes troubled with a looſeneſs; the whole ſubſtance of the Hemp being gummy, it is by no means ſurprizing that it ſhould have a reſtringent quality; and this is the reaſon why the powder of its leaves, taken in drink, is reckoned good for dyſenteries, and the duſt of the Hemp itſelf, which the labourers draw in with their breath, when they are at work upon it,

(*a*) *Tantaque vis ei eſt, ut aquæ infuſæ eam coagulare dicatur, & ideo Jumentorum alvo ſuccurrit pota in aquam.* Plin. l. xx. c. 23.

causes

causes obstructions in their lungs, and almost always, makes them asthmatic.

The root of it *(a)* boiled in water, and applied in the form of a cataplasm, softens and restores the joints of fingers or toes that are dried and shrunk. It is very good against the gout, and other humours that fall upon the nervous, muscular, and tendinous parts. It abates inflammations, dissolves tumours and hard swellings upon the joints. Beat and pounded in a mortar, with butter, when it is still fresh, it is applied to burns, which it relieves greatly when it is often renewed. The juice and decoction of it, put into the fundaments of horses, brings out the vermin that infest them.

Even the lint *(b)* which is yielded by hempen cloth, especially that which comes from the sails of vessels, is very much esteemed in physic, and the ashes of these sails serve for Spodium, Lapis Calaminaris, or Tutty.

(a) Radix contractos articulos emollit, in aqua cocta; item podagros & similes impetus: Ambustis cruda illinitur, sed sæpius mutatur priusquam arescat. Plin. l. xx. c. 23.

(b) Repertaque linteorum lanugo, è villis navium maritimarum maximè in magno usu medicinæ est, & cinis spodii vim habet. Plin. l. xix. c. 1.

End of PART I.

A TREATISE ON HEMP.

PART II.

Of the Methods of cultivating, dreſſing, and manufacturing it, as improved by the Experience of modern Times.

AFTER having laid before the reader, in the former Part, all that our reſearches could produce with reſpect to the Natural Hiſtory of Hemp, the object that appears to us moſt intereſting, is its Cultivation.

The ground intended for a Hemp-field, ought to be the very beſt that can be afforded, either near (*a*) the houſe, or along ſome ſtream,

(*a*) *At pauper rigui cuſtos alabandicus horti*
Cannabias nutrit ſilvas, quam commoda noſtro
Armamenta operi! gravis eſt tutela, ſed ipſis
Tu licet Æmonios includas retibus urſos.
Gratian. in Cyneget. v. 46.

ſtream, or water-ditch, yet ſo, that there is no ground to fear an inundation. To render this ground fruitful, we muſt not ſpare either manure or labour. Hemp-grounds muſt be dunged every year, and the better to ſecure ſucceſs, it would be proper to lay on the dung before the winter tillage, that it may waſte itſelf there, and mix the more perfectly with the ground, which, impregnated with theſe new ſalts, will derive the greater advantages from the influences of that ſeaſon, and catch more of the volatile ſalts of the air, that commonly abound moſt in the winter.

Of all the ſorts of dung, that are uſed for Hemp-grounds, pigeons-dung only, or any other kind of dung that is fully ripe, ought to be laid on before the laſt labour; as is practiſed with ſucceſs in ſeveral places. In countries where the land is ſtrong they generally lay it in heaps after Autumn. In this manner it becomes more free and light, than when it is only tilled. The ſnow and rains, which penetrate it in the winter time, and the froſts that are common in that ſeaſon, kill, if we may ſo expreſs it, that ground, as they would a chalk-ſtone, and make it ſo free, that in the month of February nothing more is wanting, but

but to lay it level by a quick and eaſy labour. All its parts, and even the moſt tender particles, will be then found extremely ſmall, light, and lively.

But different ſoils require different methods of preparation; and it is the part of men of underſtanding to diſcontinue bad cuſtoms, that have prevailed till this time, and ſubſtitute better methods in their place.

Hemp is one of thoſe plants, which Nature has not only made neceſſary, but alſo common, and ſuited to every ſort of ſoil, as well as to every climate. It is true, that countries which are extremely hot, are not favourable to it; but as this plant is but a ſhort time in the ground, if a country is at all habitable by men, we are of opinion they may alſo cultivate Hemp. Rainy ſeaſons are proper enough for ſowing it, and when it is once in condition to cover the ground, the dews alone, which are plentiful in theſe countries, will be ſufficient to bring it to maturity. It will not, to be ſure, grow ſo high as in temperate, or colder climates, but it may be, perhaps, on that very account, the more fit for uſe.

We find, by experience, that in temperate climates, *France for inſtance*, the Hemp cultivated

tivated in the ſouthern provinces has better qualities than that which grows in the northern, where the ground is fat, and not ſo warm.

In the north of America and Europe, Hemp thrives exceeding well: that commodity is exported from thence into England, Holland, and even France, to the ſhame and detriment of thoſe that cultivate it with us. Could no means be fallen upon to encourage them, and increaſe their number? What country is in better condition to apply to it, and profit by it, than France (*a*)? All her provinces produce very good Hemp; and, inſtead of taking it from ſtrangers, we ought to put ourſelves in condition to ſell it to them. Guyenne, Languedoc, Provence, Dauphiny, Auvergne, Burgundy, and Berry produce as good Hemp as can be wiſhed, and nothing is wanting but the beſt methods of cultivating and preparing it.

The firſt, and moſt important operation, ought to be performed before winter. Some do it with the plough, others with the pick-axe, and others with the ſpade; the laſt of theſe, without doubt, is the beſt, becauſe it goes

(*a*) The Council paſſed ſome Acts in December 1719, and in May and June 1722, which it would be proper to conſider.

deepeſt

deepeſt and makes the ground more free. (*a*) In the beginning of the ſpring the ground is prepared by freſh labour, for receiving the ſeed, in ſuch manner, that not ſo much as one clod be left unbroken, and the whole field be as ſmooth and in as good plight as a plot in a garden.

To have good ſeed you muſt uſe the produce of the laſt crop, and the grain muſt be clean and full grown. Seed of two years growth will not be ſo good, that of three years ſtanding will be ſtill leſs valuable; and oftentimes will not ſpring up at all. You muſt neither ſow too thin, nor too thick (*b*); both exceſſes have inconveniences that inſeparably attend them. Yet the danger of ſowing too thick is the greater. For, beſides the loſs of the ſeed, that might have been ſaved, the ground being drained of a great part of its juices, while the ſeed is ſpringing up and getting out of the earth, will not have enough left to bring it to perfection; by this means a great many ſtalks, eſpecially thoſe that are

(*a*) *Deinde utiliſſima funibus Cannabis ſeritur à favonio.* Plin. l. xix. c. 9.

(*b*) General rules cannot be given on this head; much depends on the ground and on the ſeed; it is certain, however, that it is ſown thicker than corn.

lateſt

lateſt in ſpringing, are quite choaked (*a*); or if this ſhould not be the caſe, yet ſtill they languiſh for want of nouriſhment, and the Hemp produced has neither the length nor ſtrength it would have acquired, had it been ſown thinner.

The ſeaſon for ſowing it begins not much before the firſt of April (*b*), and does not laſt beyond the end of June. The diverſity of ſoils in the ſame province, and the inconſtancy of ſeaſons, occaſion this difference. And ſo long a ſeaſon for ſowing is the more neceſſary, that it gives an opportunity to ſow the ſame field a ſecond or third time, when, by reaſon of various accidents, the firſt ſeed is loſt. But after all, the Hemp that is firſt ſown generally thrives beſt, and appears moſt beautiful, unleſs it is nipt by the froſt, or ſpoiled by the heat, when it begins to ſpring and grow up. The day on which this plant ſprings, and a few that follow it, are generally moſt critical; but, in a ſhort time, it acquires ſtrength enough to bear all the croſs accidents that can befall it: A little rain before or after the Hemp is ſown, turns much to its advantage.

(*a*) *Quo denſior eo tenuior.* Plin. loco cit.

(*b*) *Hoc tempore Cannabum ſeris.* Palladius, l. iii. c. 6.

When

When the ſeed is ſown, it muſt be put under ground, either by means of the harrow, if the field has been tilled with a plow, or with the rake, if it has been done by hand; but, however well the ſeed may be covered, you muſt not loſe ſight of your field till the Hemp gets fairly above ground. The birds, eſpecially the pigeons, are enemies, which muſt be continually kept at a diſtance. Though they do not ſcratch, nor do the leaſt injury, to corn newly ſown, when the ſeed is well covered, yet they are ſtill dreadful to Hemp-ſeed, which riſes quite out of the ground when it ſprings; whereas all other ſorts of grain lie concealed in it; and therefore the pigeons, perceiving the Hemp-ſeed at a diſtance when it riſes and diſcovers itſelf, pick it up, and all is loſt. This is almoſt the only attention which Hemp-grounds require, from the ſeed-time to the harveſt. Thoſe which are ſituated on the ſides of ſtreams or rivers, or ſurrounded by ditches, may be watered in time of great drought. In countries where their ſituation will allow it, they are drenched by letting the water run in upon them. This labour and attention in the perſon who cultivates Hemp, often turns out to his advantage, and is well

 rewarded.

rewarded. When the ground is ſown too thin, or by any accident the graſs gets up and injures the Hemp, the ſuperfluous ſtalks, or weeds, muſt be carefully pulled up, for fear they ſhould be prejudicial to the reſt.

Towards the end of July, the ſtalks which bear the flower, and are improperly denominated *female*, begin to grow yellow at the top, and white at the root, the flowers fall, the leaves wither, and this is generally the ſign of their maturity. They are then pulled up (*a*), and made up in ſmall bundles, which are ranged upon the verge of the field, taking care, as much as poſſible, to place the ſtalks, which are of the ſame length, equal at both ends, but eſpecially at the roots. It muſt alſo be obſerved, in pulling up theſe ſtalks, not to hurt thoſe that are to ſtand and bear the ſeed. This done with proper caution, will give new ſtrength to the plant that is left in the ground. For this kind of weeding not only delivers the Hemp-ground from a great number of plants, that exhauſted its ſtrength, and injured and choaked one another, but is alſo an ope-

(*a*) There would be danger in letting them ſtand longer; for beſides that they would hurt the reſt, they would themſelves become uſeleſs, and by drying in that poſition would loſe their ſtrength and their other good qualities.

ration

ration uſeful to the reſt that remain, by raiſing and ſtirring the ground about them.

In ſome places, after binding the ſmall bundles with the worſt ſtalks of Hemp, they expoſe them to the ſun, to dry the leaves before the Hemp be watered (*a*); and when they are dry enough, they cauſe them to fall off, by ſtriking every bundle againſt the wall, or a tree, or the ground: But this does not appear to be the beſt method; becauſe, beſides that it multiplies toil and labour, it alſo expoſes the Hemp to many accidents, when the ſeaſon happens to be rainy. The water which falls upon the Hemp before it is dry, makes it of a blackiſh colour, and full of ſpots. This inconvenience might be avoided, by obſerving a method, which to us appears to deſerve the preference. When the Hemp is perfectly ripe,

(*a*) The word *Rouir*, which we tranſlate *to water*, is by ſome derived from *Ros*, *Dew*, becauſe in ſome places Hemp is expoſed to the dew, to be watered thereby. In low latinity, it was ordinary to ſay *Robiare* for *Rouir*, and *Rothorium* to ſignify the place where the Hemp is watered. *Ducange*.

In the Ordonance of the Emperor Frederic, which makes the thirty-fifth title of the third book of the Conſtitutions of Sicily, this operation is called *Cannabum maturare*, *maccrare*, *diluere*, *aqua ſubigere*.

Others pretend that the word *Rouir* is derived from the red colour which the Hemp contracts during this operation.

 for

for this is a circumſtance abſolutely neceſſary, it muſt be put into the water as ſoon as it is pulled out of the ground. It's gum, which then is, in ſome reſpect, in a ſtate of fuſion, will conſequently be the more quickly diſſolved (*a*). In this condition it will not require to be more than four days in the water: whereas when it is not watered till after it is dry, it is a matter of much greater difficulty to diſſolve it, and it muſt lie in the water eight or ten days, and ſometimes more, according to the ſeaſons. Warm water forwards the effect of watering, and cold retards it.

All that are employed in the cultivation of Hemp, know how it is commonly laid down, in order to be watered. It is covered with a little ſtraw, to keep the dirt from ſticking to it, and loaded with pieces of wood and large ſtones, or other heavy materials, that it may be always five or ſix inches below the ſurface of the water.

As the phyſical effect of watering (*b*) was not formerly enquired into, they fell into ſome

(*a*) For greater accuracy, it would be proper to cut off both the extremities of the ſtalks, but eſpecially the roots, which ſerve only to ſpoil the reſt of the Hemp.

(*b*) Many have believed that the operation of the watering was the beginning of putrefaction over all the parts of the plant, without miſtakes, the conſequences of which were not perceived. The watering of Hemp producing only a proportional diſſolution of a certain quantity of the gum, which joins all the fibres of the Hemp together, and attaches them to the ſtems; it is of ſome conſequence to obſerve where, when, and how this diſſolution is effected. The fineſt and cleareſt water is always the beſt. Some make a kind of ditch on the edge of a river, where the water, being more ſtill and warm, ferments eaſily, and penetrates more quickly the parcels of Hemp that are laid in it. When they are taken out of this ditch, it will be ſufficient to waſh them in the current of the river, which will carry off all the gum and mud that would otherwiſe cleave to them. The Hemp that is watered in rivers is always the whiteſt, and of the beſt quality. That which we are obliged to lay in ditches, pools, or reſervoirs of ſtanding water, and unfit for the purpoſe, is always of

without diſtinction, and was neceſſary in order to break the ſtems with greater eaſe: but this is without foundation: the ſtem would break as well if it was not watered at all; but the Hemp could not ſo eaſily be ſeparated from it, for the reaſons that we aſſign. In a word, ſhould the Hemp lie in the water ſome days more or leſs, the difference would not appear on the ſtem, but it would be very ſenſible, and of very great conſequence in the bark.

a bad colour and a very diſagreeable ſmell, loaded with dirt, and loſes a vaſt deal in the dreſſing.

In whatever manner this operation is performed, we know that the Hemp is ſufficiently watered, when the bark is eaſily ſeparated from the ſtem. This we find out, by drawing out, every day, a few ſtalks for trial. It would be dangerous to let the Hemp lie too long in the water; the fibres of the bark, too much divided, by an undue diſſolution of the gum, would not have ſtrength enough to ſtand the effort they muſt ſuſtain, when the Hemp is peeled or braked, and a great part of it would remain with the ſtems, and be loſt in the braking.

It is therefore neceſſary, for this very reaſon, to leave the Hemp no longer in the water than is ſufficient to ſeparate the bark from the ſtem, accurately and without loſs. The ſame precaution muſt be uſed with the Hemp that bears the fruit, and remains, for ordinary, five or ſix weeks on the ground, after the other is pulled, that it may come to perfect maturity. This delay is far from being of any prejudice to the plant, as ſome have imagined; the bark, as it ripens, acquires all that force and reſiſtance,

resiſtance, which is ſuitable to its nature, and becomes preferable, eſpecially for the conſtruction of ropes, which cannot be too ſtrong or too ſolid.

In the beginning of September (*a*), or rather, when the ſeed appears to be well formed, ripe, and ready to fall of its own accord, the Hemp is to be pulled as before, and in the ſame manner ranged in parcels. In ſome places, to bring the Hemp-ſeed to perfect maturity, and diſpoſe it to leave the huſks with greater eaſe, they make in the Hemp-ground, at proper diſtances from one another, round pits, one foot in depth, and ſix in circumference. In theſe pits they range the parcels of Hemp, quite cloſe to one another, with the tops down, and the roots aloft. In this poſition, they confine them with ſtraw ropes; then round this large ſheaf they place the ground, that was taken out of the pit, that the tops may be effectually covered. The heat of the ground, and the humidity of the leaves, bring on a kind of fermentation, which rots the capſules of the Hemp-ſeed without injuring the grain. It muſt not, however, be left long in this ſituation; for it would grow muſty, and thereby

(*a*) *Semen ejus cum eſt maturum ab æquinoxio autumni deſtringitur, & ſole, aut vento, aut fumo ſiccatur.* Plin. l. xix. c. 9.

the grain become unſerviceable fer ſeed. In other places, they ſatisfy themſelves with letting the tops of the Hemp wither, and get at the grain, by beating them upon a cloth, or in a place prepared for the purpoſe, where the ripeſt and beſt part of the grain falls down, and is reſerved for ſeed till the next year. What falls not off at this firſt operation, they recover by means of an inſtrument called by the name of *a ripple*, made in the form of a rake, on the teeth of which they comb the tops of the Hemp, in ſuch manner, that the leaves and the fruit are all pulled off together: Theſe are gathered in a heap, to ferment a little; then expoſed to the ſun; and when they are quite dry, they beat them, and ſeparate the ſeeds by fanning, or putting them through a ſieve. This ſecond grain is much inferior to the firſt; accordingly, it is only uſed for oil, or for feeding birds. In conſequence of the principles we have laid down above, we are of opinion, that it will be more proper to paſs all the heads of the Hemp through the *ripple*, as ſoon as they are brought from the field; then to ſeparate, with all poſſible care, the beſt grain from the common ſort, after they have been ſuffered to ferment a little in a heap. Immediately after this, the parcels

parcels of Hemp muſt be put in proper places to be watered, as has been already explained, and theſe operations ſhould be performed during the fine weather, with all the diſpatch that time and circumſtances will permit: Every one knows the method of drying the Hemp, when it is ſufficiently watered, and of what conſequence it is to preſerve it in a dry place, until it be thought proper to peel or brake it.

We are far from condemning the method of braking of Hemp, which is practiſed in ſeveral provinces, when it is done with all the care that is neceſſary. It is, in many caſes, even preferable to that of peeling, the inconveniencies and abuſes whereof we ſhall have occaſion ſoon to explain.

In the provinces that abound with Hemp, and where the people are laborious, they generally brake it all. For this end, it muſt be made exceeding dry, that the ſtems, once put into the brake, may not come out till they are perfectly broken, and as it were ground to pieces. The fibres of the Hemp, rubbed and bruiſed by this firſt operation, are cleared of their groſſeſt gum, divided and rendered fine and ſoft; and when this work is well perform-

ed,

ed, as we have ſeen it done (*a*), the Hemp is ſeparated from the ſtem without any kind of loſs, and the advantages, which the manufacturers derive from it are very conſiderable.

To dry the Hemp, as much as it ought to be, before it be braked, we may make uſe either of the publick or private ovens; and thoſe who take this method, know very well with what precautions it ought to be done; others dry it along a wall, at a diſtance from their houſes, or in caverns made for the purpoſe, open to the ſouth, and ſheltered from the north winds, under a rock, or only covered with dry ſtones or pieces of wood with earth upon them, according to the cuſtom or conveniience of the place.

This place, which the peaſants call *a drying-place*, is commonly nine or ten feet long, ſix or ſeven in height, and five or ſix in breadth; four feet or thereabout above the fire-place, and two from its entrance, they place three pieces of green wood, about an inch or two in thickneſs acroſs the drying-place, from one wall to the other, and there fix them: On theſe perches they lay the Hemp they mean

(*a*) In ſeveral cantons of Lower Berry, Argis, Buſancois, Azay, Martizai, &c.

to

to dry, about ſix inches deep. A careful perſon conſtantly keeps a ſmall fire of Hemp-ſtems, and is very watchful that the flame riſe not ſo high as to ſet fire to the Hemp; eſpecially when it has been ſome time in the drying-place. He alſo takes care to turn the Hemp from time to time, that it may be equally dried in all its parts, throughout its whole length and thickneſs, and lays on freſh Hemp in place of that which is dry enough to be taken away, and ſent to the Brake. We ſhall give no deſcription of this inſtrument, which is as well known to ſuch as make no uſe of it, as to thoſe that do, and in caſe of need may be eaſily brought from the places where it is uſed. It conſiſts only of two pieces of wood, may be had at a very moderate price, and a workman, who has ſeen but one model of it, will be able to make as many as may ſerve a whole province.

One need ſee Hemp braked but once, to be immediately maſter of the whole art. The man or woman who breaks (for in many places it is the work of women) takes in the left hand a handful of Hemp, and in the other the upper jaw of the brake. The Hemp is put between the two jaws, and by raiſing and letting

ting fall, ſeveral times with all his force, the jaw he has in his right hand, breaks the dry ſtems under the bark that lies round them. By moving the Hemp in this manner between the two jaws, the ſtems, ground to pieces and as it were reduced to duſt, are forced to quit the Hemp: The groſſeſt part of the gum falls down like a kind of bran, and the fineſt flies away like duſt. When the half of the handful is thus broken, he puts under the brake that part which he held in his hand, and never leaves it till the whole handful be perfectly broken. It is then ſtretched upon a table, or on the ground, and when he has about two pound weight, he makes it into a parcel, which he doubles, at the ſame time twiſting it ſlightly; and this is what is called a head of Hemp, or undreſſed ſtuff. In this manner all the lower ends of the Hemp are as well divided as the tops, and the manufacturer loſes not ſo much in hards as he would otherwiſe do. All the fibres in the hand of the perſon that holds the parcel by the middle, retain, as far as poſſible, their natural length; and this firſt preparation qualifies the Hemp much better, for the other operations of the heckle, than that of peeling. A woman may break from twenty to thirty pounds of Hemp in a day,

day, and this is a very great advantage to thoſe who cultivate this plant.

Thoſe who have time and patience enough for peeling (*a*) the Hemp, are obliged to take one ſtalk after another, to break the ſtem, and draw off the hemp, by letting it run between their fingers: This method is ſo plain, and ſo eaſy, that children can perform it with as much ſucceſs as grown perſons; the aged and the infirm apply to it with equal eaſe: Generally the evenings (*b*), or thoſe times wherein nothing elſe can be done, if ſuch times there be, are thus employed. This buſineſs is particularly ſuited to thoſe who watch the cattle; but we are of opinion, that the ſtrong and laborious, who can be at no loſs for more uſeful and profitable employments, ought not to amuſe themſelves with it.

Beſides loſs of time, and the expence that muſt be ſuſtained by thoſe that give their Hemp to be peeled, this practice is alſo attended with a great many inconveniencies to

(*a*) We are of opinion, that one ought only to peel the large ſtalks of Hemp, which could not be braked without too much difficulty and trouble.

(*b*) *Ipſa Cannabis vellitur poſt vindemiam, ac lucubrationibus decorticata purgatur.* Plin. l. xix. c. 9.

the

the buyer and to the manufacturer. The Hemp that is peeled generally retains some thick parts at the end next the root, the weight of which is profitable to the seller, but contrary to the interest of the buyer; the gum and dirt it has contracted in the thick and standing pools, where it was watered, sticks constantly to it, and raises an unwholesome dust in the shop where it is dressed, which is greatly injurious to the health of the manufacturer, as well as to his pocket.

Moreover, the peeled Hemp does not always retain its whole length: One is obliged to break the stem several times to get off the bark; thus the short Hemp is mixt with the long, and this inequality is no less prejudicial: The broken fibres that are mixt up with the rest of the parcel, only produce hards, which are of no great service. After all, both these methods may have their advantages and disadvantages, their conveniencies and inconveniencies; and men of sense, who mind œconomy, may choose that which to them appears the best, according to times, places, and other circumstances.

After having considered Hemp as the fruit of the ground, or as the produce of the sweat and

and labour of the perſon who cultivates it, it remains, that we treat of thoſe qualities which render it a conſiderable object of commerce, of its different uſes in the arts, and that almoſt endleſs variety, wherein it may be employed in many ſorts of manufactures. The prejudices we inherit from our fathers, and in particular the old methods of dreſſing Hemp, have led us into a great many errors. The beſt Hemp has been often rejected, and the weakeſt, under falſe appearances, generally preferred. The qualities of hardneſs, coarſeneſs and elaſticity, have been often unjuſtly aſcribed to the former, and our ignorance of its beſt qualities has been the only means of rendering it contemptible in our eſteem.

The diverſity of ſoils, ſeaſons, and climates, as we have already obſerved, have an influence upon the qualities of this plant, as well as upon thoſe of all the other productions of the earth. The Hemp that grows on ſtrong, grey, dry, light, and ſandy grounds, is generally the beſt; thoſe of warm and temperate climates are preferable to thoſe that are produced in cold countries. The Hemp of Bretagne, for example, is better than that of

Riga;

fig. 5.
fig. 4.
fig. 7.
fig. 10.
fig. 9.

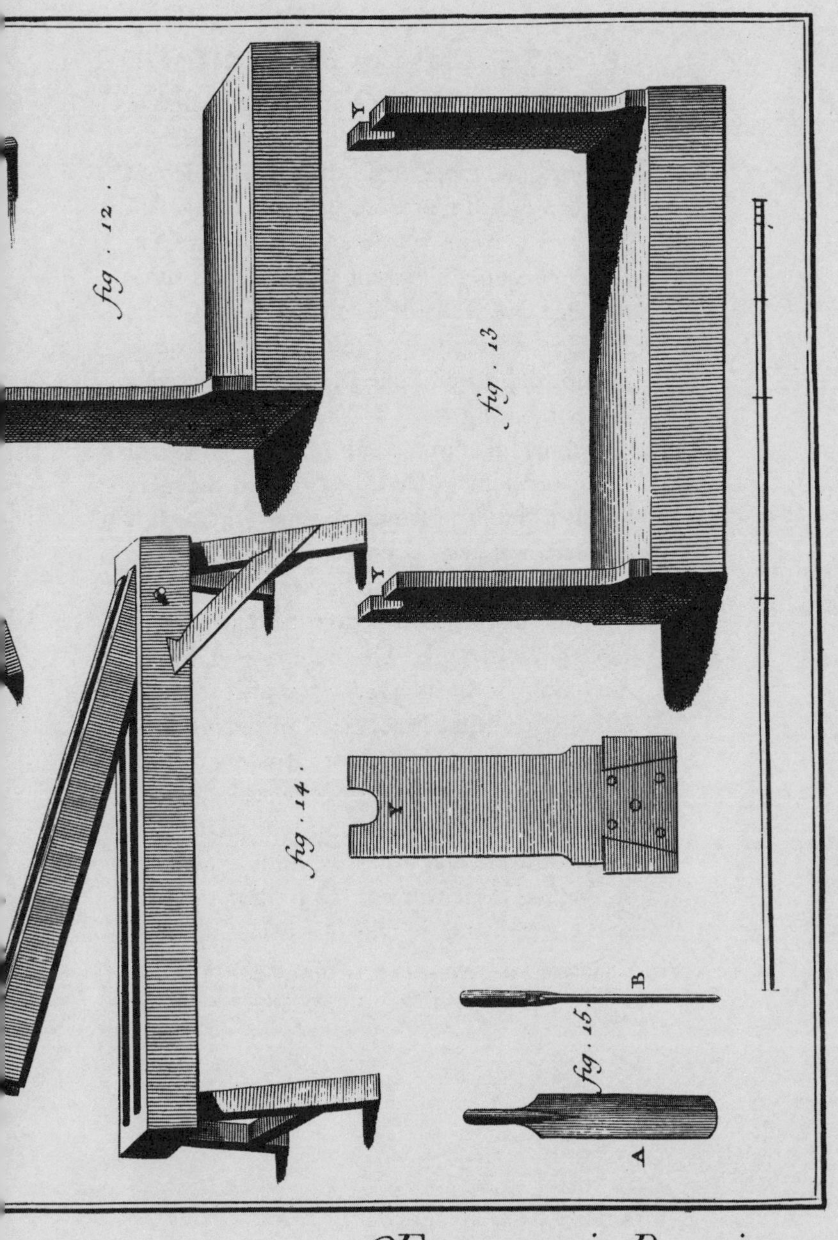

Œconomie Rustique,
Culture et Travail du Chanvre.

Riga (*a*); that of Guienne preferable to that of Bretagne; and, finally, that which is ripeſt, is, indiſputably, more uſeful than that which is too ſoon pulled; becauſe its bark being ſtill of a green colour, tender, and eaſily broken, has not acquired ſufficient ſtrength, and turns to a bad ſort of hards in the dreſſing. It is therefore of importance to thoſe who trade in Hemp, and manufacture it, to know poſitively in what ſoil it grew. It is not the colour only that ſhould determine one in their choice of this commodity; that is, very often, the mere effect of the dirty ſtanding-water in which it has been watered: Its natural colour is white, as has been demonſtrated by experiments, and the only quality which ought to be abſolutely inſiſted on, is ſtrength. This is diſcovered, by endeavouring to break a few fibres of it with the hand, when there is not time, or opportunity, to dreſs a ſample of it, before it be purchaſed. Another caution, of great importance, is, to take care that it be not damp, nor moiſt; for, beſides the loſs and waſte we muſt experience in dreſſing it, it will not fail to heat, and rot in the warehouſes in which it is laid up.

(*a*) The Hemp that grows in the Biſhoprick of Rennes, is preferred, by manufacturers, to the beſt that comes from the North.

When

When you have all poſſible aſſurance that the Hemp you are to buy is of a good ſort, you muſt examine whether the bales or packs of it are not ſophiſticated by a mixture of bad parcels of Hemp, hards, or other unprofitable ſtuff. You muſt not inſiſt upon exceeding great length in Hemp; it often happens, that the ſhort has as much reſiſtance as the long, and even ſometimes more. Of all the ſmells with which Hemp may be affected, only that of rottenneſs ought to be rejected, becauſe it is a certain mark that it is touched. And this is the greateſt defect that can be imputed to Hemp. It is, therefore, of great importance to obſerve, that it be neither rotten nor damp; the drier Hemp is, the more eaſily the gum exfoliates and parts from it; for this reaſon it is, that old Hemp, when it has other good qualities, is refined and divided more eaſily, than that which is new; ſo that, however hard or coarſe Hemp may ſeem, it ought not to be rejected, without a more circumſtantiate trial. It may not only have all the ſtrength and ſolidity that can be required, but there are alſo ways and means to procure it the ſoftneſs and flexibility neceſſary for the uſes for which it is intended.

Although the manufacturers have always, hitherto, preferred the Hemp which bears the flower, for the fabrication of thread and cloth, becauſe it is naturally ſofter, weaker, and leſs loaded with gum, than that which bears the grain, it is, however, certain, that the latter is no leſs proper for theſe purpoſes, when it is well prepared; and for the uſe of rope-yards, it has a good title to the preference. It is true, that the old method of beating, ſwingling, and heckling the Hemp, was not in condition to produce the alteration, and the effects, which are brought about by our method of preparation. As they had not then ſufficiently conſidered the conſequences of the firſt watering, they did not believe it poſſible that it could bear a ſecond; and it ſeemed to them, that if Hemp was once wetted, it could never after be of any uſe.

The ancients, whom we have, hitherto, implicitely followed, and copied in all our ordinary managements of Hemp, ſatisfied themſelves with chooſing that which had been longeſt in the water, and weakeſt, for making their fine cloth; and the longeſt and leaſt watered was only uſed for coarſe ropes, or other works

works of that kind (*a*), they imagined, that the large ribbands, which form the bark, were a kind of texture, wherein the longitudinal fibres were joined together only by a kind of little tranſverſal ones, and that theſe latter muſt be broken to obtain the ſeparation of the former; that there was no way to come at this ſeparation, but by beating, rubbing, and working the Hemp. That theſe ſuppoſed tranſverſal fibres yielded moſt eaſily to labour, being the

(*a*) Beſides the uſe formerly made of Hemp, for cloth, thread, and cordage, it was alſo the materials of a great many other works, for which there was a very great demand, ſuch as fiſhing-lines and nets, hunting-nets and gins... *Optima alabandica plagarum præcipuè uſibus*. Plin. l. xix. c. 9. Packthread, girths, ladders, bridges, trowſers, cloaths, helmets, bucklers, armour, or coats of mail, urns, baſkets, cables and tackling for ſhips, &c. as may be ſeen in Aulus Gellius, Columella, Cato, Heſychius, Pliny, Titus Livius, Xenophon, Cinegius, Pollux, Catullus, Aëtius, Paulus Æginetus, &c. Since that time, have we not ſtill extremely multiplied the uſes of it, by paper and cartoons, the conſumption of which is ſo very great? We have the greateſt ground to believe, that the impenetrability of the coats of mail, bucklers, and helmets, that were formerly made of Hemp, prepared with vinegar, proceeded from the nature of this plant, and we experience the ſame effect in paper. Is it not ſaid, that a ball, or a ſword, cannot pierce through many folds of paper joined together? ... It appears alſo, from ſeveral authors, that they often uſed raw Hemp and Flax without watering at all, *id eſt, non maceratum*, not only for making ropes, as Pliny tells us, *funes ex crudo ſparto*, but alſo for cloth, *Linteum crudarium, id eſt, ex crudo Lino, vel Cannabo factum*. Æſchylus, Pollux, Galenus, Aëtius, Paulus, Heſychius.

weakeſt, and thus the longitudinal fibres only retained their length and their ſtrength. For this end, after having bound and ſhaken, or ſwingled the Hemp, according to the cuſtoms of different countries, they put it into large mortars of wood, to be beat with wooden mallets, with an iron hoop at each end, the form and uſe whereof are now almoſt univerſally known.

In ſome places, inſtead of beating the Hemp, they make it paſs under a mill-ſtone, in a mill conſtructed like thoſe that are uſed for making oil of nuts, or Hemp-ſeed. This operation, which is commonly called *pounding the Hemp*, conſiſts in preſſing it every way, and by this action forcing the fibres to ſeparate and divide, by the exfoliation of that part of the gum which joined them together. They ſhake the Hemp, and toſs it different ways, that it may receive the various impreſſions of the mallet, or the mill, during this firſt preparation : But ſtill this was not ſufficient to qualify it for making ropes, even of the coarſeſt ſort.

It is well known, how hard and ſevere this firſt operation is to the poor work people, who are obliged to apply to it. And how much the

duſt,

duſt, they draw in with their breath, is prejudicial to their healths, and even to their lives. Yet notwithſtanding ſo much pain and fatigue, the Hemp requires ſtill another operation, called heckling, which is little leſs noxious to them. The Heckles they uſe are various, with reſpect to their ſize, form, and fineneſs, according to the difference of countries, and the beauty of the works they are intended for; but in all caſes the method of working, and the end propoſed are the ſame.

We ſhall give no deſcription of theſe Heckles (*a*); they are generally known, and may be met with every where; they may be ſeen, drawn with the utmoſt exactneſs, and in their juſt proportions, in the third volume of the Encyclopedia, page 154. in the article *Chanvre*. We will not diſown, that we have taken from it, all that we could find relating to this work of ours; and as this little treatiſe may ſpread further than that immenſe collection, we reckon it our honour, as it at the ſame time, gives us pleaſure, to ſpread the

(*a*) The Heckles uſed by the ancients, had their teeth bent in the form of hooks; whereas ours are ſtrait pointed, and ſtand perpendicular to the inſtrument. *Et ipſa tamen pectitur ferreis hamis, donec omnis membrana decorticetur.* Plin. l. xix. c. 1.

knowledge we have derived from it, though we do not always agree with the authors in ſome points. Our method inſerted in the article *Filaſſe*, ſhews to advantage, the zeal as well as the diſintereſtedneſs of the Editors, who, throughout their work, are no leſs ambitious to edify, than to enrich the public.

The buſineſs of a Heckler conſiſts, then, in ſeparating, throughout their whole length, the fibres of the Hemp, which the mill or the mallet have divided only in part. The teeth of the heckle carry off a part of the gum, which is thereby reduced to duſt; and by dreſſing and dividing, over again, the filaments into which they enter, ſeparate them intirely. The oftner this operation is repeated with different ſorts of heckles, coarſe, fine, and finer ſtill, the more the Hemp acquires of ſoftneſs, whiteneſs, and fineneſs, whether it is intended for ropes, or to be worked into cloth.

In this manner it was the antients prepared it (*a*); thus ſome prepare it to this day; and

(*a*) *Noſtro more Cannabis, aut Linum vulſum ſiccatur, in aquam mergitur, & maceratur; deinde tunditur, mox pectitur, poſtea netur, deinde texitur, textumque adhuc fuſcum eſt, donec frequentibus lotionibus, & aſperſionibus aquæ candorem ſibi conciliet. Ars eſt quippe, qua candor ille quæritur, ſed detexto tantum* &

thus cuſtoms, good or bad, are every day continued to perpetuity. In this manner the Hemp prepared for cordage, ſtill retains a hardneſs and a gum, that renders the ropes ſtiff, coarſe, and not ſo fit for uſe. What is intended to be made into cloth, produces an ill coloured thread, that is coarſe, loaded with gum, and ſuch, that it cannot be uſed without paſſing through ſeveral lyes. The cloth made of it is very hard to be bleached, and cannot be brought even to a very indifferent colour, till after ſeveral months of fatigue and labour.

We ſhall not enlarge farther upon this ancient method; numbers of experiments ſince our late diſcovery, and the reflections they have ſuggeſted, have recovered from their former prejudices, many perſons as much diſtinguiſhed by their rank as by their penetration, on whom popular errors make no impreſſion. Being convinced themſelves by the juſtneſs of their reaſoning, as well as by the experiments they have made, they have publiſhed, ſupported, and defended, the goodneſs

& jam Linteo; crudariam telam vulgò dicimus, quæ poſtquam detecta eſt lotionem lixivam, non eſt experta, ſimiliter & crudarium filum, quod à netu lixivo lavacro maceratum non fuit. Salmaſ. Exercit. Plin. pag. 765.

of our method, against the bigotry of the vulgar, who are not able, by their own sagacity, to perceive all the consequences of it. They have demonstrated, that the fibres of Hemp have as much occasion to be washed and purified from their gum, to make good thread and fine cloth, as the finest wool to be cleaned and purged of the sweat that cleaves to it, in order to be spun, and undergo the necessary preparations to its being made into fine stuffs. This is what, to this present time, has been quite unknown, and is to be our principal object in the remaining part of our work.

After having long considered the various means that might be found to relieve those who work upon Hemp, and observed, in the Hemp itself, those admirable qualities, of which, hitherto, no improvement has been made, we found that the common *watering* of Hemp was nothing but the dissolution of a tenacious gum, natural to the plant, the parts whereof are joined together merely by means of it; and that, in order to this first preparation, it was sufficient to leave the Hemp in the water, in proportion to the quantity and tenacity of this gum, that after having prepared it only for being peeled, or *bra-*

ked,

ked, it ſeemed very proper to give it a ſecond watering, to ſoften the bark, that ſtill remains hard, elaſtic, and incapable of being brought to a proper degree of fineneſs. Accordingly, by the different experiments we made, in the preſence, and under the direction of Monſieur Dodart, Intendant of Berry, we have found means, eaſily, and without expence, to give it thoſe qualities that are natural to it, and the uſes whereof were not hitherto known.

The water that has already had power to ſeparate the bark from the ſtem, ſerves alſo to divide, without trouble or hazard, the fibres from one another, by a total diſſolution of the gum that remains in them. For this purpoſe, the Hemp intended to be put into the water, is divided into ſmall parcels; theſe are taken by the middle and laid double, twiſting them ſlightly, or tying them ſoftly with a piece of ſtrong packthread, that ſo they may be ſtirred and managed eaſily without mixing.

After they have all imbibed the water, they muſt be put in a veſſel of wood, or ſtone, in the ſame manner as thread are ſtowed in a lye-tub. The veſſel is then filled with water, in which the Hemp is left, for ſome days, to moiſten, as far as is neceſſary for diſſolving the

the gum. Three or four days are ſufficient for this operation; and if one has leiſure to preſs every parcel of Hemp, to ſtir it and work it often in the water, which it would be neceſſary alſo to renew, this diſſolution might be attained more expeditiouſly, and twenty-four or thirty hours would, in that caſe, be ſufficient for the operation.

When you ſee that the Hemp is ſufficiently penetrated by the water, and cleared from the coarſeſt part of its gum, it muſt be taken out by ſingle parcels, wrung, and waſhed in ſome river, to purge it, as much as poſſible, from the muddy and gummy water, that remains in it. After it is thus cleared, it muſt be beat upon a board, to divide farther the parts that may ſtill remain too groſs; then you ſtretch out, upon an upright piece of ſtrong, ſolid wood, every ſingle parcel of wet Hemp, after having looſed the packthread (*a*). Then you ſtrike it lengthways, with the edge of the inſtrument that laundreſſes uſe for their linen, till the thick parts at the ends are ſufficiently divided. The parcels muſt not be beat too

(*a*) We have found by experience, that it was more convenient not to uſe packthread in this caſe, provided care be taken to twiſt the Hemp, ſo that the fibres may not be mixed or diſordered.

much;

much; for the fibres, by that means, being too much ſeparated and weakened, will not have ſtrength enough to ſtand the operations of the heckle; and this is a caution the neceſſity and conſequences whereof can only be known by experience. There is even good ground to think, that by leaving the Hemp long enough in the water to obtain the diviſion of the fibres, by the diſſolution of the gum only, we might diſpenſe with the beating it altogether; but the different qualities of Hemp would require ſuch particular attention, that it would not be adviſable to take this method. The more quickly the operation is performed, the Hemp runs the leſs danger; for there is ſome reaſon to believe, that, by lying too long in the water, it might have its fibres intirely diſſolved, and reduced to pure gum. This obſervation leads to a great many remarks upon cordage, hempen cloth, and paper, which it might be tedious to inſert particularly here.

After this eaſy labour, which, after all, is the longeſt that is neceſſary, every parcel, one after another, muſt be waſhed over again in running water, and then it will appear what ſucceſs is to be expected from this method.

All

All the fibres of the Hemp, thus beat, are divided in the water, washed, disengaged from one another, and seem to be as completely dressed as if they had already passed through the heckle. The more rapid, clear, and beautiful the water is, the more are these fibres bleached and purified. When the Hemp appears clear enough, and totally purged from its dirt and nastiness, we take it out of the water, wring it, open it, and expose it to the air, then lay it on a pole in the sun to drip and dry.

We might also use, for a second watering, the common lyes of ashes (*a*), either by making those lyes for the purpose, or taking advantage of those which are made frequently for linen-cloth. From the different experiments we have made, and the observations of many persons, who with the same assiduity have applied themselves to this matter, we have discovered, that the gum of Hemp, which has been pretty well cleared before, is by no means unfriendly to linen-cloth when it is mixed with it; and that it will be sufficient,

(*a*) Hemp of a green grassy colour, may be brought, by the use of lye, to its greatest perfection, and the labour of beating it may be almost intirely dispensed with; it becomes white without trouble or loss.

in

in ſuch a caſe, to put only a layer of clean ſtraw, about two inches thick, in the bottom of the lye-tub, in order to filtrate and purify the water, and to attract all the mud and gum that is in it. By this eaſy precaution, the ſalts of the lye, thus ſet at liberty, exert their whole activity upon the Hemp, or linen, which is penetrated by the water; and it has never been obſerved, that it left any ſpot, or blemiſh. It will be eaſily imagined, that the warmth of the water, and the alkali of the aſhes, muſt operate a diſſolution much more expeditiouſly than can be effected by cold water; but it will be no leſs neceſſary to beat the Hemp, that may ſtill remain not ſufficiently divided, and to waſh it, at leaſt for the laſt time, in clear running water, to purge it intirely from the lye water and its gum.

Beſides the two methods which have already been approved and practiſed, in ſeveral provinces of the kingdom, we have diſcovered, that the operations neceſſary for the bleaching of Hemp may be ſtill very much abridged. The various objections and queries, that ſeveral perſons have offered upon our memorial, the execution whereof appears to them difficult and inconvenient, have obliged us to

ſhew,

fig. 4.
fig. 6
R
R
R
R
S
fig. 3.
fig. 2.
fig. 1.

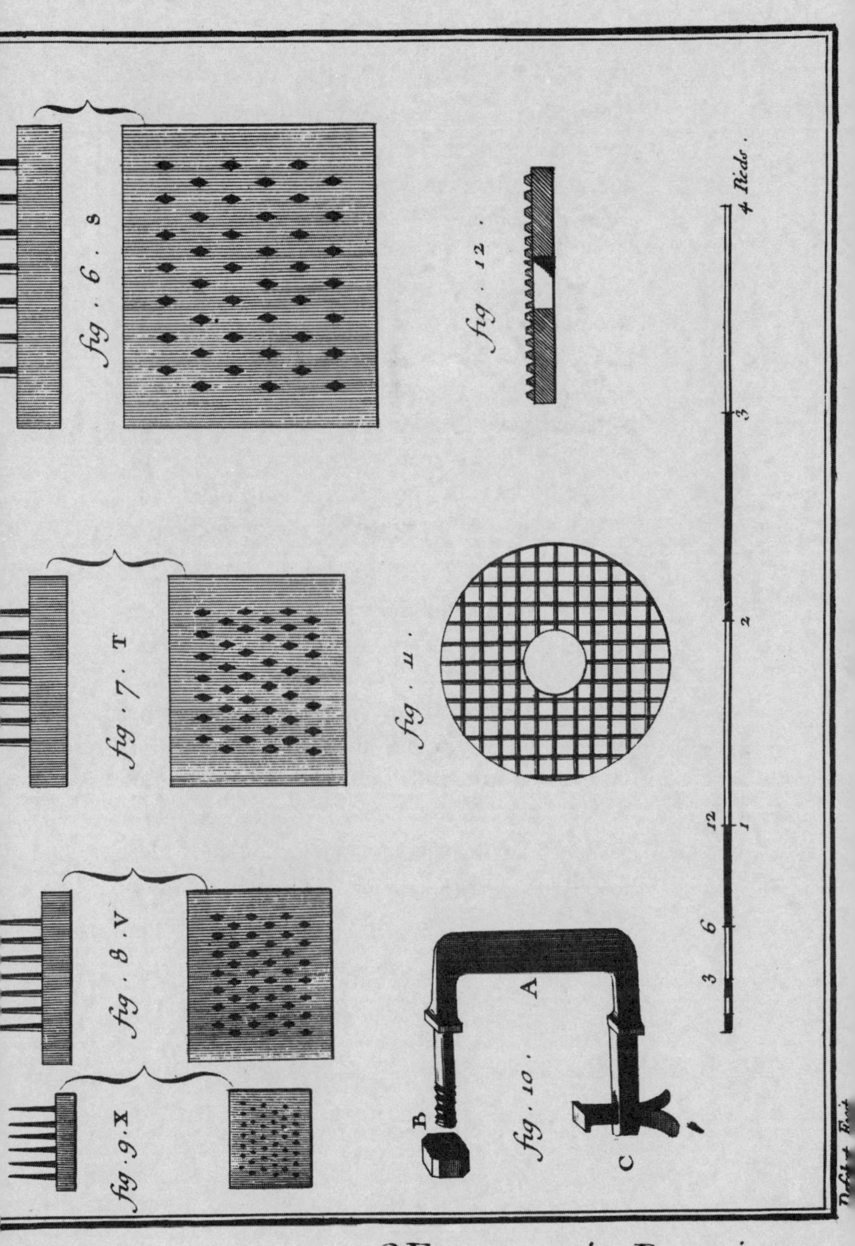

OEconomie Rustique,

Culture et Travail du Chanvre.

ſhew, that if it is not an eaſy matter to lay down upon paper the plaineſt and moſt ſimple operations, it is at leaſt very eaſy to make them intelligible, by performing them, but once, in their preſence. We made it appear, in ſeveral cities of Berry, that no more than two hours time is neceſſary to waſh and bleach Hemp, in winter as well as in ſummer; eſpecially if we have the advantage of ſprings, the water whereof is commonly warm in the winter (*a*). In this manner we have, in about twelve hours time, produced Hemp bleached, prepared, and ſpun, with all the perfection it was capable of.

Heat being abſolutely neceſſary to diſſolve the gum, of which you want to clear your Hemp, it is moſt proper to wait till you have fair and mild weather, not to diſcourage nor diſtreſs your workmen, who will think that a hard work, which will oblige them conſtantly to put their hands in cold icy water, or perhaps, for that reaſon, neglect ſome of the operations eſſentially neceſſary, in order to ſucceſs.

(*a*) For the ſame reaſon, river water is more proper in ſummer than that of ſprings, which, at that time of the year, is too cold.

Thus,

Thus, thoſe who would expeditiouſly make experiments upon two or three pounds of Hemp, may divide them into ſeveral little parcels of three ounces each, or thereabouts. Theſe they are to put into a proper quantity of water, as hot as they can bear their hand in it. There they are to leave them to moiſten and ſouple about half an hour; then take out each parcel by itſelf, wring it, dip it, ſqueeze it, and work it in the water, as laundreſſes do linen cloaths, when they have ſoaped them, ſo as to prevent their mixing together, and being hurtful to one another. After this firſt operation, the water, being dirty, thick, and loaded with gum, muſt be thrown away. You muſt then take a ſecond water, as hot as the former, in which the Hemp muſt be worked as before; then a third, until the Hemp appears to be clear enough. After theſe three bathings, if any ribbands of bark ſtill remain that are too broad, or not enough divided, you muſt take the inſtrument and beat them moderately, to procure a further diviſion.

This operation being compleated, you muſt bring your Hemp to ſome river, and waſh it in the ſtream, which will then carry away all the

the rest of the gum. In this manner, the fibres of the Hemp, like so many threads of silk, are disengaged, divided, purified, refined, and whitened; because the gum, which is the sole principle of their union, is also the cause of their nastiness, and of the different colours to be seen on Hemp; finally, take it out of the water to drip and dry, as has been said above.

When the Hemp is dry enough, it must be doubled with caution, twisting it, at the same time, slightly, to keep the threads from mixing together, and thus it is given to the Hemp-dresser, to prepare it for spinning (*a*). It will

(*a*) Let none tell me, that the operations above described, are too tedious, too expensive, and too painful. Let them say, rather, as they are new, they have not got the habit of doing them; for if we compare them with the care, the pains, and the expence that people bear patiently, by custom, *for instance, when they prepare their bread at home*, we shall see they can give no other reason but custom. Let us consider the loss of time those are at, who go to the market to buy corn, then to the mill, to which they are often obliged to go several times, and wait their turn; to bring the flour back to their own houses, and there bestow all the time, which the present method of making bread requires, whether in town or country: It is certain that this takes up a considerable space of time, and diverts those that are employed in it, from their other business: It also is attended with expence, and often occasions irregularities. If we can give credit to an estimate made by a very able man, the whole of the loss occasioned by buying the corn,

not be neceſſary to beat it ſo long as before. This work, formerly ſo hard on account of the ſtrength it required, and ſo dangerous on account of the fatal duſt the workman drew in with his breath, will be, henceforth, only a buſineſs moderately ſevere. There will be no occaſion to look for machines to ſave the labour of men, and prevent the dangers attending the work. The buſineſs of a hemp-dreſſer, henceforward, would be confined to an eaſy beating of the Hemp, and the common operations of the heckle; it is ſo much the more eaſy, that the materials are ſofter, and no longer exhale unwholeſome duſt; and, more-

corn, and the baking of bread in private houſes, amounts, every year, to forty-one million ſix hundred and ſixty-ſix thouſand five hundred and fifty livres, (or 1,822,911l. 3s. 9d. ſterling,) which the ſtate would ſave if no bread were prepared but by the baker. We do not include, however, in this loſs, the expence of wood, which is very conſiderable, when every private family is obliged to heat its own oven; neither do we include here the private loſs which reſults from the badneſs of the bread in compariſon of that which would be well prepared, well baked, and eaten in ſeaſon, when it would be uſeful for the body, and conducive to health. But, notwithſtanding that the alterations I propoſe in the operations upon Hemp, are advantageous, they are rejected, they are neglected; thoſe I have juſt now mentioned, with reſpect to bread, are deſtructive, they are received with ſubmiſſion, and obſerved; ſo powerful is the force of habit and prejudice, on the vulgar and the ignorant.

over, there is ſcarce any waſte in this operation. The ſecond beating ſerves only to divide a ſecond time the fibres of the Hemp that reunite in drying, and this renders it white, ſmooth, flexible, ſoft, and fit to receive all the preparations of the heckle. If you want it to paſs through the fineſt heckle, the Hemp, thus waſhed, will afford dreſſed ſtuff capable of producing the beſt thread, comparable to what is yielded by the fineſt flax, and you will have little more than a third of very good hards.

Now this hards, that was formerly an object of diſcouragement, and ſold commonly to rope-makers for two ſols and ſix deniers; (that is, about five farthings the pound;) by this new operation becomes a matter of very great advantage. By carding them like wool, they produce a fine, marrowy, and white ſubſtance, the true uſe whereof was never diſcovered till now. It may not only be uſed alone, as it is, for making of wadding, which, in many reſpects, will have the advantage of the ordinary ſort; but, moreover, it may be ſpun, and made into very beautiful thread (*a*). It

(*a*) There is no queſtion, but thread made of ſuch hards, would perfectly ſuit wax and tallow-chandlers for wicks, &c.

may

may be alſo mixed with cotton and ſilk, with wool, and even with hair; and the thread, that reſults from theſe different mixtures, affords, by its vaſt variety, materials for new eſſays, very intereſting to the arts, and of vaſt utility to ſeveral ſorts of manufactures.

The Hemp thus prepared, may be alſo dyed (*a*) like ſilk, wool, or raw cotton, either red, blue, yellow, &c. or other colours, ſuitable to the uſes intended to be made of it. It will receive and retain, with great eaſe, the tinctures that are given it, in order to be made into ſtuffs, cloth, ſtockings, and garments of all kinds; or even into tapeſtries, embroideries, and other ſorts of furniture.

The principal advantage that Hemp, intended for theſe uſes, will have over wool, grogram yarn, and cotton, is, that it may be uſed without ſpinning, or even combing. It

(*a*) Pliny informs us, that in his time, they had got the art of dying Hempen-cloth as well as woollen; that Alexander the Great, in his expedition againſt the Indians, to ſurprize them, cauſed the ſails of his ſhips and flags to be dyed. The ſails of the veſſel, in which Cleopatra made her eſcape with Marc Antony, at Cape Figo, in Albania, were dyed purple. The Hempen-cloth, that ſerved to cover the ſtreets, publick places, and theatres, was dyed red, blue, or other expenſive colours, according to the magnificence and opulence of thoſe, who gave publick entertainments.

will be in no danger from thoſe worms, which commonly eat woollen cloth; and the beauty, as well as the laſting nature and low price of it, will render it preferable to any other material. The different trials, that have been made of this ſort, leave no room to doubt of ſucceſs in other attempts of the kind.

Mixtures that may be made with Hemp, will be the more valuable, as they will leſſen, more than one half, the price of the moſt expenſive and uncommon materials, with which its hards are incorporated. In a word, we ſhall have the advantage and ſatisfaction of finding in a plant, which grows commonly among us, the means of indemnifying ourſelves for a part of thoſe productions which we are obliged to fetch, every day, from foreign and very diſtant countries, or even of diſpenſing with them altogether (*a*).

(*a*) In Hemp, duly prepared, beſides ſilk, hair, wool, and cotton, we find a new material, which, till now, had never any exiſtence in commerce, nor in manufactures, but may be made extremely uſeful in both. By mixing it with wool, for inſtance, half and half, we work it into caps and cloths, which are no ways different from thoſe that are made of wool alone, even in their greateſt perfection. By mixing the hards of it with cotton, we make cloth and ſtuffs, and even coverlids, which, with regard to whiteneſs, ſoftneſs, fineneſs, and other qualities, may be preferred to thoſe, which, at firſt, we only propoſe to imitate.

Nor

We have already given, in ſeveral cities of the kingdom, ſuch ſamples of theſe mixtures, as have been demanded of us, that the firſt ſight of them has excited the ſurprize, and merited the approbation of the beſt judges.

We have not yet near exhauſted all the combinations which may multiply the uſe of Hemp, in its different forms. The cloth that is made of Hemp, thus prepared, will not be ſo

Nor is it to be doubted, but it may be alſo employed in manufactures of hats, becauſe it will be an eaſy matter to make felts of it, if we mix it with wool proper for that purpoſe. We may alſo make it into hunting waiſtcoats, and waiſtcoats for the army; breeches, hunting bags, carpets for gaming and writing tables, and a great many other works, generally made of leather, and very expenſive.

In a word, by the different combinations made with the hards of Hemp, they aſſume the nature and properties of the materials, with which they are mixed, and the price of ſuch materials will be leſſened in proportion as they are now dear, uncommon, and rare.

We ſhall not, at preſent, be more particular upon this ſubject; theſe obſervations, being in our hands only, will be always confined and imperfect; but the ideas we have ſuggeſted, may eaſily receive a better form, and be carried further, by the knowledge and experience of thoſe who are at the head of commerce and manufactures. All that we can aſſure our readers, is, that by introducing into trade and manufactures, as a fifth material, prepared Hemp, which may be conſidered as a new creation, we ſhall not only add a fifth to the four that have been long known; but, to uſe a *mathematical expreſſion*, we ſhall by this means raiſe each of them, far above the *fifth power*.

long in bleaching (*a*), and even the thread (*b*) will not have occaſion to paſs through ſo many lyes as were formerly found neceſſary. The ſails of ſhips will not be ſo ſtiff and heavy, the ropes will be found more flexible and ſtrong, and move with more eaſe and expedition.

Theſe firſt diſcoveries have led us to think, that even the waſte, the coarſeſt dreſſings of the Hemp, and the ſweepings of the ſhops where it is prepared, contain ſomething valuable, which, in former times, was either thrown into the fire, or upon the dunghill, becauſe its uſe was not known. Yet after all, it wants only to be beat, cleaned, and purified by water, in order to be of excellent ſervice in paper manufactories (*c*). The experience

(*a*) We find in Pliny, that formerly in bleaching cloth they uſed a ſort of poppy. . . *Eſt & inter Papavera, genus quoddam quo candorem Lintea præcipue trahunt.* Plin. l. xix. c. 1. If we have no acquaintance with that ſort of poppy, which, Pliny informs us, was generally uſed in bleaching Hempen-cloth, we ſhall be indemnified for that loſs, by employing, for the ſame purpoſe, Horſe-cheſnuts, the preparations whereof are neither expenſive nor troubleſome. As this fruit is every where to be met with, all may have opportunities of taking a tryal of it, and we ſhall ſhew the method of uſing it at the end of this work.

(*b*) *At veteres Cannabem aut Linum iterum in filo polibant, & filici crebre illidebant cum aquâ, textumque rurſus clavis tundebant.* Salmaſ. Exercit. Plin. pag. 765.

(*c*) There is a paper-mill near Toulon, which for a long time has made uſe of the junck of ropes.

we have had of this, ſhews evidently, that it might be made a matter of great importance.

After the particular account we have given of the nature and properties of Hemp, we doubt not but the people of the country will avail themſelves of all the advantages they may attain by the practice of theſe new methods. If they apply themſelves to the cultivation of Hemp, and carry to perfection the methods of preparing it, what reſources will they not find, in employments (*a*) ſo profita-

(*a*) *Quæſivit Lanam & Linum, & operata eſt conſilio manuum ſuarum. . .* In former times, among the Hebrews, the Greeks, the Romans, and almoſt all other nations, women only were employed in making linen, or hempen-cloth and ſtuffs. Even Queens and Princeſſes were not aſhamed of theſe employments, which we abandon to the meaneſt of all artiſts. We ſee here Bathſheba, the wife of David, applying, very earneſtly, to every particular of houſhold affairs, making herſelf, or cauſing her maids to make, under her own eye, linen and woollen cloth, for cloathing her family. Queen Penelope, the wife of Ulyſſes, wove, with her own hands, a very fine web. The Goddeſs Calypſo is deſcribed to us, as buſied in the ſame manner. Omphale, the Queen of Lydia, applied alſo to ſpinning; and did not the famous Hercules, to pleaſe her, like an effeminate man, take the diſtaff and the ſpindle? Alexander the Great, ſpeaking to the mother of Darius, and ſhewing her the garment he had upon him, addreſſed her thus: "Mother, you ſee a garment that was made by the hands of my ſiſters; it is not only a preſent they have made me, but alſo the work of their hands." Auguſtus, for ordinary, in his family, uſed no cloaths, but what were

ble, and at the ſame time ſo eaſy? For to conſider only its moſt common qualities, it muſt be acknowledged, that it is a commodity abſolutely neceſſary. The uſe of it extends to almoſt all the purpoſes of commerce and of life. There is no ſtate nor condition that can be without it. The very perſon who cultivates it, is the firſt to make uſe of it for cloathing; and of all his labours, this is often the only fruit which he retains. He cultivates it through neceſſity, and through neceſſity he keeps it. There is a ſingular kind of circulation in this commodity; nothing that bears a near reſemblance to it is to be found in the other productions of life. The more it is uſed, the more the cultivation muſt increaſe; and the more you cultivate it, the more you increaſe its conſumption. The cultivation alone is

made by the Empreſs, his wife, his ſiſter, his daughters, and his grand-daughters.

Eginhard tells us almoſt the ſame thing of Charlemagne. Telemachus told his mother, who wanted to concern herſelf with other matters: "Mother, apply to your proper works; take the diſtaff, weave your web, and give orders to your maids."

Sozomen obſerves of women, that they ſallied out of the place, where they were weaving their webs, and, armed only with their ſhuttles, ſtuck them into ſome holy martyrs at Gaza, in Palæſtine. Bibl. de Calmet, Prov. c. xxxi. ver. 13.

is a labour that requires inhabitants, and the consumption of it serves to maintain them. In the different methods of preparing it, young men and women, old men and children, find employment, in proportion to their strength and ability. Some find business in preparing the ground, and sowing it; others pull the Hemp, and peel it; others make ropes or cloth; all of them join in the consumption, and make use of it; and every one jointly and severally contributes to renew their own work, while they are relieving their wants.

The manufacture of Hemp, then, is one that is naturally suited to the country; as it is necessary to all, it should spread every where. The manufacturer, in the proper season, becomes cultivator; and the cultivator, with all his family, in his turn, becomes manufacturer, as soon as he has finished his harvest. Then the different preparations which Hemp requires, give him an opportunity to avail himself of the time wherein, on account of the severity or inconstancy of the weather, he would have otherwise nothing to do.

He may, at the same time, have continual employment for those that are not able to apply to the business of husbandry. Hence arises

a general

a general advantage to the whole country, and a low price at the firſt hand. The whole buſineſs is performed with œconomy, and without prejudice to other affairs in the family, or about the farm. The Hemp being prepared, preſerved, and diſtributed, with precaution and care, is a ſure means to procure what materials are wanted, at a low price. Hence, the goods manufactured from Hemp may be afforded cheap, and conſequently are ſure of a vent. He that ſells only the ſuperfluities of his time and his labour, ſells his work much cheaper than the perſon that has no other reſource. Why do the Indians ſell at ſixteen or twenty ſols per yard, the printed cloths which our company traders ſell to us again at fifty or ſixty? Becauſe the people in theſe countries, being at little or no expence for their victuals and cloathing, think themſelves very happy to find vent for their works of cotton, which they would often ſee rotting in their houſes, if the low price at the firſt hand did not facilitate the ſale of them. Is it not alſo known, that the goods manufactured in Swiſſerland come to a good account, and are diſperſed through a great part of Europe, becauſe that people, accuſtomed to a hard and laborious life, are ſatisfied with a moderate profit, to ſecure

cure themſelves good vent for their commodities, and continual employment. If they make not their cuſtomers pay dear for their time, they know, on the other hand, that they do not loſe it (*a*).

But not to go too far, for examples of this diſperſed manufacture, which enriches ſome of our own provinces as well as other places, let us take a view of Flanders, Picardy, Normandy, Bretagne, &c. where the manufactures of linen are the ornament, as well as the riches of the whole country. Such manufactures,

(*a*) The prodigious demand for the manufactures of Swiſſerland, Sileſia, &c. is perhaps owing to the liberty they enjoy in theſe countries to ſell, promiſcuouſly, all ſorts of linen, without being obliged to repair to ſtamp offices. Though theſe offices have been ſet up in France, with a very good intention, yet there are inconveniencies that often attend them, as they keep the manufactures from being multiplied and extended. Every manufacturer is obliged to get all his pieces of linen ſtamped, whence it follows that linen cannot be made, but in the neighbourhood of thoſe places where ſuch offices are eſtabliſhed; ſo that, on pretence of improving induſtry, and ſecuring better fabricks, the manufactures are often weakened and confined: Hence it is, that notwithſtanding the great demand there is for our linen, our fabricks, for theſe twenty five or thirty years paſt, have not increaſed ſo much as they might have done. They continue ſtill confined to the places where they firſt began; while thoſe of our neighbours have made a progreſs, that might juſtly excite our emulation and ambition.

without

without doubt, ought to be encouraged and protected. The arts, disperséd in the country, render it populous and fruitful (*a*). There the Prince, in time of need, finds soldiers and artists, and the earth has people to cultivate it; but we complain daily that we see the country abandoned. The indigence and misery, to which a great part of its inhabitants are reduced for want of work and food, oblige them to take refuge in the cities, carrying with them numerous families, which, soon after, are disperséd and come to nothing.

The manufactures, on the contrary, that are settled in the country (*b*) continue there, find support, and multiply the inhabitants. What hurry of business? what circulation

(*a*) It is pretended, that no more than fifteen acres of land are necessary to employ and maintain a farmer, that has ten mouths to provide for. The ancient Romans gave only seven to a peasant of the most numerous family.

(*b*) *Sindonem fecit & vendidit, cingulum tradidit Chananeo...* The Merchants of Phœnicia, designed by the word *Chananeo*, traded to all parts of the world; and the neighbourhood of their country to that of Judæa, gave the virtuous woman an opportunity to sell whatever she chused to profit by. She sold her own works, and those of her servants; this trade was neither mean, nor disgraceful; the most honourable men, and even Princes and Kings, did not at all scruple to deal in it. "The virtuous woman sold her fine linen, her cloth, &c." Bibl de Calm. Pro. c. xxxi. v. 24.

may

may we not ſee in thoſe places? The peaſant, together with the fruits of the earth, brings the produce of his induſtry to market. The merchant makes up his aſſortment of goods, and fills his warehouſes, without quitting his cloſet or his ſhop, and the manufacturing-farmer is as ſure of a vent for his work as for his corn or his pulſe (*a*). Nothing, perhaps, is wanting to animate a little thoſe countries, where no manufactures are as yet eſtabliſhed, but ſome abatement of the taxes to thoſe who diſtinguiſh themſelves; a diſtribution of prizes to thoſe who merit them;. or proper gratifications for every piece of goods that ſhould be manufactured; in a word, any ſort of encouragement to thoſe who ſhould deſerve it. This is the way that manufactures have been multiplied and brought to perfection in *Stotland* and *Ireland*; and by ſuch means they begin to get footing in *Breſſe*. Upon the whole, why ſhould we give to ſtrangers that

(*a*) The commerce of farmers in the country, merits the particular attention of politicians. If induſtry is not encouraged, we may have ſome towns in a flouriſhing ſtate by means of their manufactures; but the body of the nation will be ſtill in a bad ſituation. The greateſt part of the people will live in miſery, or ſcarce live at all; and, to get any ſupport from them, they muſt be oppreſſed in a moſt barbarous manner. *Princip. de Negot. par M. l'Abbé de Mably*. 1757. *pag*. 236.

profit

profit which might be kept in the heart of our own kingdom, by forming ſuch eſtabliſhments as cannot fail to augment our population and riches (*a*)? Why take, for inſtance, from Bruſſels or Germany, our coarſe checks and ticking, the conſumption whereof is ſo conſiderable, and the manufacture ſo eaſy. The true intereſt of a ſtate is not always to inſiſt upon too uncommon perfection in its fabricks.

The manufacture for which there is the greateſt demand, according to the principles of the Dutch, is always that which ought to be encouraged, in preference to any other. They are not ſo fond of great perfection, as great conſumption (*b*).

And, indeed, what ſignifies it how we beſtow our Hemps, if they give us employment, and find a vent. It would be always proper to draw foreign manufactures into France; but we ought not, out of fondneſs for novelty,

(*a*) The King of Spain, intending to encourage the manufactures of his own kingdom, which he eſtabliſhed ſome years ago, has juſt now prohibited all ſorts of ſtuffs, manufactured at Genoa, to be imported into any of his eſtates, which will greatly hurt the trade of that republick.

(*b*) Great perfection naturally follows great conſumption, and the employment of many hands.

which

which is objected to the French, neglect thoſe manufactures that are naturally moſt ſuited to the climate (*a*). It has not, perhaps, been ſaid, without reaſon, that the commerce of France had carried the principles of M. Colbert to an extravagant height, by multiplying prodigiouſly all the different manufactures that are ſettled in towns, without taking ſufficient notice of thoſe that ought to be diſperſed through the country. There is ſtill ground to fear, that we have been too guilty of ſacrificing to the arts of luxury, the eſſential buſineſs of agriculture, and the neceſſary manufactures, which, like that of Hemp, are inſeparably connected with it.

It is not only true that Hemp, from its own nature, ought to be the object of a manufacture diſperſed through the country; but we are alſo of opinion, that it can never, with any advantage, be the buſineſs of a manufacture crowded into a town. None can be ignorant of the inconveniencies that attend theſe

(*a*) The policy of the Engliſh is admirable in this reſpect: To ſecure the conſumption of their wool, of which they have much greater plenty than of Hemp, and to encourage their manufactures, they have prohibited the burying of the dead in linen, and they have a manufacture of woollen ſtuff, intended for that purpoſe, and for that only.

crowded

crowded manufactures; the expence of setting them up, and the room they take, the expence of keeping up the buildings and that of the administration, which is, almost always, faulty, the roguery of the greatest part of the manufacturers, and their cabals, sometimes also the avarice and inattention of the proprietors, suggest but too many reasons to reject the ideas of this kind that might be formed. The manufactures of hempen and linen cloth are less in condition to support this charge, than any other. The facility with which the country people, in their hovels, might apply to the several preparations of Hemp, and to carry hempen cloth to perfection, would give them a profit much superior to that of a manufacture crowded into towns, and the inequality of the competition between these two must surely ruin the latter.

Few commodities but iron, chrystal, gunpowder, sugar, glass, porcelane, tapestry, and some others of that sort, can bear the expence of manufactures crowded into towns; because the merchandize and fabricks of these kinds yield a profit far superior, and much more constant, than manufactures of hempen and linen cloth.

A manu-

A manufacture dispersed in the country, is therefore the only one, that is by nature suited to fabricks of Hemp and linen. The use of these is too necessary and too common, the fabricks too extensive, too simple, and too well known to be an advantageous prospect for a great enterprize; and much more can be promised from their dispersion, than their reunion into one place. A dispersed manufacture is liable to no expence, and requires little or no money in advance; it insinuates itself wherever there are laborious or idle hands, and, by its natural relation to agriculture, it necessarily, and in a peculiar manner, concurs with it, in increasing population, encouraging the cultivators, enriching the provinces, and rendering the whole kingdom happy, flourishing, and powerful.

Omne tulit punctum qui miscuit utile dulci.

End of PART II.

The Method of Bleaching Linen, Hempen-cloth, and Stuffs with HORSE-CHESNUTS.

THE eagerneſs of ſeveral Provinces to know our new method of preparing Hemp, advertiſed in ſeveral Journals and periodical Pamphlets, and actually put in practice, with ſucceſs, in ſeveral cities of the kingdom, gives us ground to believe, that the new diſcovery we have made, not long ago (*a*), with regard to the uſe and properties of the Horſe-cheſnut, will give no leſs ſatisfaction. After divers experiments relating to our primary object, and repeated obſervations on this tree and its fruit, we have diſcovered, that the Horſe-cheſnut abounds with aſtringent, aluminous, deterſive, lixivial, and ſaponaceous juices, the uſe whereof may be made extremely advantageous to men, with reſpect to Phyſic, and the Improvement of Arts; and as the Bleaching of Linen, Hemp-cloth, and Stuffs

(*a*) In September 1757.

appears

appears to have a natural connection with the operations we have recommended in the preceding treatiſe, we are of opinion that it will be proper not to ſeparate them.

The method of proceeding is very plain, and is here ſubjoined. It will be ſufficient, in this caſe, to peel the Horſe-cheſnuts, and to raſp them with a ſugar-raſp, into a proper quantity of cold water (*a*). Rain or river water is beſt. The juice they yield, diſſolved and

(*a*) The Horſe-cheſnuts muſt be raſped very fine, and the water prepared ten or twelve hours before it be uſed, that it may be the better impregnated in the juice of this fruit. It is to be ſtirred from time to time: but in order to uſe it, it muſt be taken off the dregs with a ſkimmer, or, by inclining the veſſel, poured off, about half a quarter of an hour after it has been ſtirred, while it continues white and appears like ſoap-water, it foams and crackles in the ſame manner. The operation of raſping the Cheſnuts, after once accuſtomed to it, will not be a matter ſo tedious nor difficult, as it appears at firſt. We are very ſenſible, that when a great quantity of theſe Cheſnuts is to be uſed, the labour muſt be abridged, and rendered more ſimple, which may be done, by putting the Cheſnuts, when dry and peeled, into a mill, and reducing them to a very fine powder; and if they are not dry enough, they may be formed into a paſte, which will eaſily diſſolve in water.

It has been proved by experience, that this Horſe-cheſnut water may be uſed inſtead of ſoap for ſcouring cloth, after having looſed the dirt with fullers earth, as is practiſed in manufactories and fulling-mills. For this purpoſe you muſt put into the veſſel a quantity of Horſe-cheſnut water ſufficient to moiſten and foment the cloth you want to be fulled. This water ſhould have a proportionable degree of heat, and care muſt be taken to renew it, as often as it ſhall be found neceſſary.

 mixed

mixed with a proportionable quantity of water, is very proper for washing, cleaning, and bleaching Hempen-cloth and Stuffs. A score of Chesnuts will require about ten or twelve French pints (or five or six English wine gallons) of water.

To make use of it, the water must be heated till you cannot bear your hand in it. If you cannot absolutely dispense with soap, you must, at least, take less of it than usual, rubbing with it only those places where the dirt is most tenacious, and the saving will be considerable, as the expence to those who are obliged every day to use soap in their business, such as laundresses, and fullers of stockings and stuffs, is very great.

We have got stockings and woollen-caps fulled with Horse-chesnut water only, and they have perfectly taken the dye: We have also made trial of this water upon stuffs fulled in a fulling-mill, and met with the same success. The linen or hemp-cloth bleached with this water, retains something of a bluish gloss, which is no wise disagreeable, especially when, after having cleaned it two or three times with Horse-chesnut water, you take care to wash it well in river water. A vast number of experiments,

periments, performed in our preſence, in the ſeveral towns of Berry, ſerve to confirm our firſt eſſay, and give full ſatisfaction to thoſe who mind ſuch matters. But what has fully convinced us of the connection between this latter and our former diſcovery, is the proof we have had of it, particularly upon Hemp, which we put into the Horſe-cheſnut water, to ſteep and macerate for ſome days. After being ſlightly rubbed in it, the fibres of the Hemp were divided, ſoftened, and became much whiter than thoſe which had been only waſhed with pure water. The activity of the ſalts, wherewith the Horſe-cheſnut abounds, and the oil it contains, carried off intirely from the Hemp, the gum that ſtuck to it moſt obſtinately, and when it could not totally diſſolve, it forced to exfoliate.

Let none, after all, imagine, that this water produces upon linen, hempen-cloth, or ſtuffs, ſuch a ſenſible effect as the beſt kind of ſoap; but this method of bleaching at leaſt requires no expence. The weakeſt children may peel and raſp Horſe-cheſnuts without fear of committing miſtakes; and when, by repeated waſhings, they have drawn out all the juice, the paſte, which will remain without bitter-

bitterneſs, and almoſt quite inſipid, mixed with bran, may ſerve for food to the fowls and other animals in the yard. And, to conclude, even the aſhes of the Horſe-cheſnuts will make very good lyes.

This firſt diſcovery, however uſeful it may be, is only a ſketch of the operations reſulting from the firſt obſervation. Far from flattering ourſelves that we have exhauſted the ſubject, on which many experiments ſtill remain to be made, we hope that ſuch happy beginnings will excite perſons of greater ability, and of a better underſtanding, to enquire further into the other qualities of the Horſe-cheſnut tree, and of its fruit, as well as the diverſity of uſes to which both may be applied.

As to the phyſical properties of the Horſe-cheſnut, which we have made it our buſineſs to obſerve and point out, it was rather to recommend the enquiry to others, than to make it our principal ſtudy. It is known, that Horſe-cheſnut, reduced to powder, has a powerful effect in provoking to ſneeze, and muſt not be uſed without great precaution. As the Horſe-cheſnuts contain a deal of aluminous juices, they ſeem proper for curing of hæmorrhages, either by an infuſion of them mixed

with

with drink, or by fumigation. It is, probably, on account of their aſtringent qualities, that the farriers cauſe horſes to ſwallow them for a cough. It is alſo imagined, that the aluminous juices, which abound in the Horſe-cheſnut tree, render it a difficult matter to burn it, and it yields but very little aſhes, becauſe it is exceeding full of pores.

FINIS.

ACKNOWLEDGEMENTS

This facsimile is from the original copy in the Herbarium of Sir Joseph Banks, the noted economic botanist; with the kind permission of the British Museum Press.

The two illustrations from the Encyclopédie (1751–1772), 'Landmark of the 18C', are mentioned by Marcandier on p. 53 (the hemp brake, or horse, at fig. 11). They are included both for convenience and as a tribute to his main inspiration; by the kind assistance of the late Jean Gimpel and his wife Catherine.

A
TREATISE
ON
HEMP.

PART III.
POSTSCRIPT

Science without conscience is but the ruination of the soul. RABELAIS

For me, Marcandier's magisterial *Treatise on Hemp* (Cannabis sativa L.) is also a *tour de force* of natural economics, imbued with the enlightened principles of that new school of social philosophers known as the *Economistes*, or *Physiocrats*, which flourished under their leader Quesnay during the 1760's of pre-industrial France.

These men are honoured as "the most exciting and the most contemporary group of economists in the whole history of economic thought".[1] "The most exciting,

[1] *The Economics of Physiocracy*. R. L. Meek. London. 1962.

because the birth of Physiocracy was in fact the birth of the science of economics. The most contemporary, because the Physiocrats' major preoccupations, in both the theoretical and the practical fields, were strikingly similar to those of present day economists. Relying on natural laws, this school attempted to bring prosperity to France by formulating new theories and policies on the economy".[2]

"Of all the natural physical laws, none was more vital than that which pointed to the soil as being the unique source of wealth. For Physiocrats, manufactures were useful, even necessary; but they merely gave natural wealth or materials a different shape".[3]

As "an important source of medicine, food, clothing, shelter and communication",[4] the economics of hemp are indeed both exciting and contemporary. Never more so than today, now that

[2] *The Encyclopédistes as Individuals.* F. A. and S. L. Kafker. OUP, 1988.
[3] *Ency. Brit.* 1961.
[4] *Ecologist/Hempathy.* 1980.

global warming is at last recognised by governments, and the ravages of rampant consumer-industrialism have finally to be addressed.

The real wealth and health of society, as Marcandier and his fellow *Physiocrats* so ably demonstrated, was not to be gained from government-led mercantilism, slave-trading and emerging factory systems for processing imported raw materials (in rivalry perhaps with England), but by favouring a national subsistence economy based on agriculture and rural industry. In effect, the greatest good for the greatest number, for which true hemp culture is the great exemplar.

Even Smith's fateful apologia for industrialism explicitly allows that "capital employed in agriculture not only puts in motion a greater quantity of productive labour than any equal capital employed in manufactures, but also, in proportion to the quantity of productive labour it employs, it adds a much greater value to the annual produce of the land and labour of the country, while it encreases the real wealth and revenue of its inhabitants".[5] Indeed, Smith had met the Physiocrats, through Hume in Paris, and acknowledged his debt to them in his Preface to the above: "their new system was the nearest approximation to the truth yet published on political economy". (His mathematical bent, no doubt, was the cause of his 'chopping-up' of the workman into fractions, in order to effect his brute theory of the division of labour.)

Evidence that physiocratic principles were still held down to recent times may be found in the Development Commission Act of 1909. Known today as the Rural Development Commission (RDC), it was formed by Parliament in order to promote the economic development of Great Britain, with the specific remit to revive the hemp and flax industry. Professor Vargas Eyre of Wye Agricultural College was appointed adviser. By 1913, he reported that it was both feasible and desirable. His contemporary A. D. Hall agreed, since "hemp had always yielded excellent results

[5] *Wealth of Nations.* Adam Smith. London, 1776.

for the small man".[6] They stressed the need for common centres where small farmers could take their hemp and flax for processing. (Watermills could once more provide the ideal centres for co-operative rural industries.) More crops, they said, suited to small farmers were urgently required.[7]

Sadly, World War One 'took-off' so many small farmers, and the worthy RDC have yet to realise their hemp and flax remit. Since then the prejudice, misinformation, ignorance and Mumbo-Jumbo which has smothered hemp's historic usefulness, have their origins in the US Marihuana Tax Act of 1937. It represents the ultimate industrial enclosure which ushered in this Age of Synthetics. The corrupt and secretive passage of this law, foisted on the American people by the capital-industrialist corporate members of the 1934 American Liberty League, who waged treason and war against President Roosevelt and the New Deal,[8] is the

[6] *A Pilgrimage of British Farming.* London, 1913.
[7] *Journal of the Royal Agricultural Society of England.* 1913.
[8] *Du Pont Dynasty: Behind the Nylon Curtain.* G. Colby. US, 1974/84.

root and provenance of our Misuse of Drugs Act, 1971. We now know that this infamous act was primarily intended to protect the billion-dollar wood-pulp paper industry (and its associates) from the rising competition of home-grown hemp under the US national recovery.[9]

"If a law be bad it is one thing to oppose the practice of it" (i.e. to advocate or break it), "but it is quite a different thing to expose its errors, to reason on its defects, and to show cause how it should be repealed, or why another ought to be substituted in its place".[10]

Now it is true that hemp has, at last, been partially restored here—just after this **treefree**® hemp-content paper was launched in early 1993. That is commendable, that is a triumph for the trees; for trees are the chief supporters of Nature.[11]

[9] *Unravelling the American Dilemma: The Demonization of Marihuana.* John C. Lupien. US, 1995.
[10] *Rights of Man.* Thomas Paine. London, 1792.
[11] *Triumph of the Tree.* J. S. Collis. London, 1950. ("the clear-eyed prophecies of woe carried through from Cobbett to Carlyle and from Ruskin to William Morris were stifled by the Fabians, who suddenly and calmly *accepted industrialism.*")

But the global chainsaw massacre of transnational companies continues; with 385 million acres of forest destroyed during the 1980's and only one per cent replaced[12], and with around 27,000 species now estimated to become extinct each year.[13]

Consumption of paper, largely wood-pulp since only the mid-19th century, has increased some 20 times this century to about 300 million tonnes per annum, globally. Governments must now combine environmental principle with hemp practice. Annual indigenous plant fibres—true hemp, cornstraws or flax—in widespread, small-scale cultivation, will provide up to ten times as much fibre per acre than trees annually, and from their wastes or 'hurds' alone—to say nothing of the prime line fibre. This is both the key to rural 'low impact development'[14] and national

[12] *Forests for Life.* Observer/WWF UK. 1996.

[13] *Corporate Power, Corruption and the Destruction of the World's Forests. EIA, 1996.*

[14] *Low Impact Development.* Simon Fairlie (details on the back cover). Permaculture? "The beds were intersected by furrows; in each trench there stood, as if on guard, ranks of hemp stalks, the cypresses of the vegetable

subsistence. Let us start with allotment land and free seed for the unemployed, as provided for in the 1931 Agricultural Act; then the redundant county small-holdings; initiate land reforms, and tax-breaks for woodless, home-grown paper, cloth, nets, ropes, etc. With "hemp education, manufacture and pharmacy"[15] the order of the day: dig hemp for victory,[16] the natural way.

Vive Marcandier! *Floreat cannabis.* The plant, the whole plant, and nothing but the plant.

John Hanson/Cht.
2nd. October, 1996.

garden, calm, straight and green; their leaves and their scents served to defend the beds, for through their leaves no serpent dares to pass, and their scent kills insects and caterpillars." *Pan Tadeusz.* Adam Michiewicz. Paris 1934.

[15] The objects of my family trust, Cht. (1976–1996).

[16] *Hemp for Victory.* U.S. Dept. Agriculture. 1941. This exhortatory documentary restored the truth about hemp, when Pearl Harbour cut off imported fibres from Manila and the East Indies. By the war's end, however, 'the nylon curtain' descended and the Marihuana Tax Act was re-imposed.

H E M P.

definitions from the

Oxford English Dictionary

2nd. Ed. 1989

by kind permission of the
Oxford University Press

Oxonìa Erat Demonstrandum Cannabis.

CHTHONIUS

hemp (hɛmp), *sb.* Forms: 1 hænep, henep, 4– hemp, (4–7 hempe, 6 hemppe). [OE. *hęnep, hænep* = OLG. **hanap*, **hanip*, MDu. and Du. *hennep*, LG. *hemp*, OHG. *hanaf, -if, -uf* (MHG. *hanef*, Ger. *hanf*), ON. *hampr* (Sw. *hampa*, Da. *hamp*):—OTeut. **hanpi-z*, **hanapi-z*, cogn. with Gr. κάνναβις, L. *cannabis*: cf. also Lith. *kanapés*, OSlav. *konoplja*, Pers. *kanab*. The word is perh. not Aryan, but adopted in Greek, Germanic, etc. from some common source.]

1. An annual herbaceous plant, *Cannabis sativa*, N.O. *Urticaceæ*, a native of Western and Central Asia, cultivated for its valuable fibre.

It is a diœcious plant, of which the female is more vigorous and long-lived than the male, whence the sexes were popularly mistaken, and the female called *carl* or *winter h.*, the male *fimble* (i.e. female), *barren*, or *summer h.*: see CARL HEMP and FIMBLE.

(The quotations from the *Saxon Leechdoms* appear to refer to some wild British plant, perh. the *wild hemp* of 5.)

a **1000** *Gloss.* in Wr.-Wülcker *Voc.* 198/12 *Cannabum*, hænep. *Ibid.* 198/15 *Cannabin*, hænep. *c* **1000** *Sax. Leechd.* I. 16 Herba chamepitys þæt is henep [*v.r.* hænep]. *Ibid.* 228 Ðeos wyrt þe man cannane silfatica, & oþrum naman henep nemneþ. *c* **1325** [implied in HEMPSEED]. *c* **1440** *Promp. Parv.* 235/2 Hempe, *canabum.* **1523** FITZHERB. *Husb.* §146 In Marche is tyme to sowe flaxe & hempe. **1551** TURNER *Herbal* I. Hjb, Hempe..is profitable for many thynges..and specially to make stronge cables, and roopes of. **1578** LYTE *Dodoens* I. l. 72 Hempe is called in Greeke κάνναβις..in English Hempe, Neckeweede, and Gallow-grasse. **1794** MARTYN *Rousseau's Bot.* xxix. 456 Hemp has a five parted calyx in the flowers which bear stamens, but in the pistilliferous ones it is one-leaved, entire and gaping on the side. **1883** *Harper's Mag.* Oct. 715/2 Land that will grow hemp will grow anything.

b. **1523**, etc. [see CARL HEMP]. **1577**, etc. [see FIMBLE]. **1597** GERARDE *Herbal* II. ccxxxviii. (1633) 709 The male is called Charle Hempe and Winter Hempe. The female Barren Hempe, and Sommer Hempe. **1753** CHAMBERS *Cycl. Supp.* s.v., The male Hemp, or summer Hemp, which bears no seeds, and is called by the farmers *Fimble-hemp*, will have its stalks turn white in July. *Ibid.*, The remaining plants, which are the female Hemp, called by the farmer *Karle-hemp*, are to be left till Michaelmas.

2. The cortical fibre of this plant, used for making cordage, and woven into stout fabrics.

c **1300** *Havelok* 782 Hemp to maken of gode lines And stronge ropes to his netes. ? *a* **1366** CHAUCER *Rom. Rose* 1233 A sukkenye, That not of hempe ne [? hempene] heerdis was. **1404** *Nottingham Rec.* II. 22, xlv. strykes de hempe, iiijd. **1550** CROWLEY *Epigr.* 1139 Newe halters of hemppe. **1634** SIR T. HERBERT *Trav.* 105 Long, deepe prams, sowed together with hempe and cord. **1662–3** PEPYS *Diary* 18 Feb., Casting up..accounts of 500 tons of hemp brought from Riga. **1722** SEWEL *Hist. Quakers* VII. (1795) II. 10 Committed to Bridewell and required to beat hemp. **1881** *Daily News* 18 Apr. 2/8 Tows and hemps move off very freely.

3. In allusion to a rope for hanging.

† *stretchhemp*, a person worthy of the gallows. † *to wag hemp*, to be hanged.

1532 MORE *Confut. Tindale* Wks. 715/1 To mocke the sacrament the blessed body of god, and ful like a stretch hempe, call it but cake bred. *Ibid.*, Tindall..feareth not (like one yᵗ would at length wagge hempe in the winde) to mocke at all such miracles. **1599** SHAKS. *Hen. V*, III. vi. 45 Let not Hempe his Wind-pipe suffocate. **1654** WHITLOCK *Zootomia* 60 Of no small use to purge a Common-wealth, without the expence of Hemp. **1849** JAMES *Woodman* xxviii, If his people catch me, I shall taste hemp. **1864** LOWELL *Fireside Trav.* 56 [He] express[ed] a desire for instant hemp rather than listen to any more ghostly consolations.

b. (See quot.) Cf. HEMPY *sb.*

1785 GROSE *Dict. Vulg. T.* s.v., *Young hemp*, an appellation for a graceless boy.

4. A narcotic drug obtained from the resinous exudation of the Indian hemp; bhang; hashish.

1870 YEATS *Nat. Hist. Comm.* 195 Hemp is employed in other forms besides churrus as a narcotic. **1893** *Nation* (N.Y.) 9 Feb. 108/1 Its votaries have taken to opium and hemp, the latter of which Sir Lepel Griffin says is far more injurious than tobacco.

5. With qualifying words, applied to numerous other plants yielding a useful fibre, or otherwise resembling hemp: as **African hemp**, (*a*) = *bowstring hemp* (*a*); (*b*) *Sparmannia africana* (Miller *Plant-names*). **American false** (*Laportea*) *Canadensis* and *U. cannab* (Miller). **Manilla h.**, the fibre of *Musa textilis*, the Banana family. **mountain h.**, *Hyoscyam insanus* (Syd. Soc. Lex.). **nettle h.** = HEM NETTLE. **Peruvian h.**, *Bonapartea junc* **Queensland h.**, the tropical weed *S. rhombifolia* (N.O. *Malvaceæ*), called also Pad or Native Lucerne, and Jelly Leaf. **ramie** *Bœhmeria nivea*. **sisal h.**, the fibre of species *Agave*, esp. *A. Sisalana*. **Virginian h.**, **will h.**, *Acnida cannabina*, an amarantaceous ma plant, native of eastern U.S. **water h.**, a na given to *Eupatorium cannabinum* and *Bid tripartita*, in U.S. to *Acnida cannabina*. **wild** *Eupatorium cannabinum* (Gerarde), a *Galeopsis Tetrahit* (Britten & Holland).

1597 GERARDE *Herbal* II. ccxxviii. 573 This wilde Her called *Cannabis spuria*, and also *Cannabina Spuria*, bastarde Hempe. *Ibid.*, In English wilde hempe, Ne hempe, bastard hempe. *Ibid.* II. ccxxix. 574 The bastard wilde Hempes, especially those of the water, are ca commonly *Hepatorium Cannabinum*..in English, w Hempe, bastard and water Agrimonie. **1611** COT *Chanure sauvage*, Bastard Hempe, wild Hempe, Ne Hempe. **1688** R. HOLME *Armoury* II. 72/2 The bas Hemp is with several Burs, or hairy Knobs at a distance the stalk. **1744** J. WILSON *Synops. Brit. Pl.* 95 *Lam cannabino folio vulgare*..Nettle Hemp, or rather Her leav'd dead Nettle. **1796** WITHERING *Brit. Plants* (ed. *Bidens tripartita*, Trifid Doubletooth, Water Hemp, W Agrimony. **1866** *Treas. Bot.* 350/2 *Crotalaria juncea*..T plant is extensively cultivated in..India, on account of valuable fibre yielded by its inner bark, which is known the names of Sunn-hemp, Bombay Hemp, Madras He Brown Hemp, etc. *Ibid.* 1015/2 The Bowstring Hemps.. stemless perennial plants. **1897** MORRIS *Austral Engl. Queensland Hemp*...is not endemic in Australia.

6. *attrib.* and *Comb.*

a. *attrib.* Of hemp; made of hemp, hemp

a **1400–50** *Alexander* 2224 Oure pepill..Halis vp he cordis. **1549** *Privy Council Acts* II. 349/1 Hemp ropes, weight. **1599** *Acc. Bk. W. Wray* in *Antiquary* XXXII. A p[air] of hempe shetes. **1630** B. JONSON *New Inn* I. iii, may, perhaps, take a degree at Tiburne..And so goe for Laureat in hempe circle! **1662–3** PEPYS *Diary* 24 F Captn. Cocke and I upon his hemp accounts till 9 at ni **1668** T. THOMPSON *Eng. Rogue* II. i, You have no rem against a hemp halter I hope. **1875** R. F. MARTIN tr. *Ha Winding Mach.* 32 The wires..in each strand must twisted round a hemp core. **1893** *Daily News* 2 Mar. Inquiry..into the trade in all preparations of hemp drug Bengal.

b. Comb., as ***hemp-close***, ***-cock***, ***-gar*** ***-hammer***, ***-harvest***, ***-harvester***, ***-heck*** ***-knocker***, ***-plant***, ***-plot***, ***-ridge***, ***-seller***, ***-smok*** ***-spinner***, ***-stalk***, ***-top***; ***hemp-leaved***, ***-li*** ***-packed***, ***-producing*** adjs.; **hemp-beater**, person employed in beating the rotted stems hemp, so as to detach the fibre; an instrum used in doing this; **hemp-brake**, an instrum for bruising or breaking hemp; **hemp-bush**, Australian Malvaceous plant, *Plagiantl pulchellus*, yielding a hemp-like fibre; **hem cake**, the residue of crushed hempseed, af **h.**, *Datisca hirta* (Miller *Ibid.*). **bastard h.**, na given to the British plants Hemp-nettle a Hemp Agrimony (Britten & Holland). **Ben h.**, **Bombay h.**, **Madras h.**, *Crotalaria jun* (Miller). **bowstring h.**, (*a*) a plant of the ger *Sanseviera*, esp. *S. guineensis*, a liliaceous pl of tropical Africa, the leaf-fibres of which used by the natives for bowstrings and making ropes; (*b*) in India, *S. Roxburghia* also *Calatropis gigantea* (N.O. *Asclepiadace* **brown Indian h.**, *Hibiscus cannabinus* (Mille **Canada** or **Indian h.**, *Apocynum cannabinum* N. American perennial (J. Smith *Dict. Ec Pl.*). **Cretan h.**, *Datisca cannabina* (Miller). **h h.**, an old name for *Galeopsis Ladanum* (Mille **Indian h.**, a tropical variety of Common Hen *Cannabis Indica*. **jute** or **plant h.**, *Corcho capsularis* (Encycl. Brit.). **Kentucky h.**, *Urt*

:traction of the oil; **hemp-dike, -dub, -pit** *ial.*), a small pond for steeping green hemp; **:mp-hards, -hurds**: see HARDS; **hemp-atcheler, -heckler** = HEMP-DRESSER; **hemp-l**, the oil pressed out of hempseed; **hemp-alm**, a palm, *Chamærops excelsa*, of China and pan, the fibres of which are made into ordage; †**hemp-roll** (see quot.); **hemp-sick** *a.* f. HEMPEN 1 b, quot. 1785); **hempwort**, any ant of the Hemp family; **hemp-yard**, a piece ground on which hemp is grown, a hemp-rth or hemp-close.

1615 E. S. *Brit. Buss* in Arb. *Garner* III. 653 Will convert our vagabonds..into lusty *hempbeaters. **1725** VANBR. *ov. Wife* IV. iii, That fist of her's will make an admirable mp-beater [in Bridewell]. **1886** *Syd. Soc. Lex.* s.v., :mpbeaters, carders, and spinners..suffer from dust sing from the material. **1873** BOUTELL & AVELING *raldry* Gloss., **Hemp-brake* or *Hackle*, an instrument for uising hemp. **1878** *Ure's Dict. Arts* IV. 364 *Hemp cake chiefly used for adulterating linseed cake. **1698** FROGER *y.* 58 The Fields..are like those of our *Hemp-Closes. **69** WORLIDGE *Syst. Agric.* xii. (1681) 250 Stick them on tops of *Hemp-cocks or Wheat-sheaves. **1877–89** *N.W. nc. Gloss.*, **Hemp-croft, -garth, -yard*, the gardens ached to old cottages commonly went by one of these mes, as they were in former days used mainly for growing mp. **1878** *Cumberld. Gloss.* **Hemp dub*, a small pond used steeping green hemp. **1627** *Merton Reg.* II. 296 Unum empegarth simul cum libertate communii. **1663** *MS. lent. of Barlby* (Yorksh.), An orchard, a hemp-garth, two rdens. **1637** NABBES *Microcosm.* v, The shrieks of mented ghosts [are] nothing to the noise of *hemp-mmers. **1707** MORTIMER *Husb.* v. xi. 120 'Tis a very great p to the Poor; the *Hemp-harvest coming after the other rvest. **1724–7** RAMSAY *Tea-t. Misc.*, *Bob of Dumblane*, nd me your braw *hemp heckle. **1579** LANGHAM *Gard. alth* (1633) 300 Apply it with *Hempe-hurds to the heate the Liuer and stomach. **1586** *Praise of Mus.* 76 That petie d counterfait Musick which..*hemp-knockers [make] w[t] ir beetles. **1744** *Hemp-leaved [see sense 5]. **1712** tr. *met's Hist. Drugs* I. 158 The burnt Oil they make use of *Hemp-Oil. **1839** R. S. ROBINSON *Naut. Steam Eng.* 39 is kind..keeps steam-tight with far less friction than the :mp-packed piston. **16..** *Add. MS.* 31028 lf. 7 (N.W. ic. Gloss.) Drowned in a *hempe pitt near a little sink of npe. **1832** G. A. HERKLOTS tr. *Cust. Moosulm. India* oss., *Gunja*..the leaves or young leaf-buds of the *hemp nt. **1678** BUTLER *Hud.* III. ii. 43 Like Thieves that in a emp-plot lie Secur'd against the Hue and Cry. **1824** ACTAGGART *Gallovid. Encycl.*, **Hemp-riggs*, ridges of fat d whereon hemp was sown in the olden time. **1696** J. F. *rchant's Ware-ho.* 23 The next..Linnen, is called emp Roles, it is always brought into England brown, and strong coarse Linnen..and..when whited very good for eets for Poor People. **1785** *Life Miss Davis* 5 He..was ivicted and hanged..and her *hemp-sick husband laid in earth. **1875** KNIGHT *Dict. Mech.* 1099/2 *Hemp-stalks beaten to remove the bark and cellular pith from the er. **1853–5** *Cassell's Pop. Educ.* IV. 29/1 Cannabinaceæ or empworts. **1378** *Durham MS. Cell. Roll*, In plumbo pto pro uno aqueducto in le *Hempyard. **1725** BRADLEY *m. Dict.* s.v. *Hemp*, Pigeons dung is good for Hemp rds.

Hence **hemp** *v. trans.* (*rare*), to halter, to hang. 1659 CLEVELAND *Lenten Litany* II. i, That if it please thee assist Our Agitators and their List, And Hemp them with entle twist.

:mp-dresser. One who hackles hemp.

1659 CLEVELAND *Times* 81 No zealous Hemp-dresser yet p'd me in The Laver of Adoption from my Sin. **1723** *id. Gaz.* No. 6171/10 Benjamin Bellamy..Hempdresser.

b. *pl.* The name of a kind of country-dance.

756 AMORY *J. Buncle* (1770) II. 25 We..had the hemp-ssers one night, which is, you know..the most difficult, l laborious of all the country dances. **1827** in Hone *Every- Bk.* II. 122, I have 'footed it' away in Sir Roger de verley, the hemp-dressers, &c.

mpen ('hɛmpən), *a.* (*sb.*) Also 4–5 **hempyn(e, :ne**, (6–7 **hempton**, 7 **hemton**), 6–8 **hemping**. [f. MP *sb.* + -EN[4]. Not recorded in OE.; but cf. HG. *hanafîn* (Ger. *hänfen*), LG. *hempen*.]

1. Made of hemp; of or pertaining to hemp. *hempen homespun*, homespun cloth made of hemp; ice, one clad in such cloth, one of rustic and coarse nners.

375 BARBOUR *Bruce* x. 360 He gert sym of the ledows.. hempyn rapis ledderis ma. **1398** TREVISA *Barth. De P.R.* lxii. (1495) 898 The weke is made of hempen threde. *c* **1440** [see HEMPY *a.* 1]. **1535** LATIMER *Serm. Insurr. in North* (1844) 29 It is no knot of an hempton girdle. **1558** PHAER *Æneid* v. 552 But [he] hyt the hemping corde, and of the knot the bands he brast. **1590** SHAKS. *Mids. N.* III. i. 79 What hempen home-spuns haue we swaggering here? **1651** *Miller of Mansf.* 8 Good browne hempton sheetes. **1669** WORLIDGE *Syst. Agric.* (1681) 44 A very great succour to the poor, the Hempen Harvest coming after other Harvests. **1703** *Wakes Colne, Essex, Overseers' Acc.* (MS.), 6 yards of hempinge cloth for two shifts for Suzan Beets. **1776** ADAM SMITH *W.N.* I. x. II. (1869) I. 128 Weavers of linen and hempen cloth. **1887** BOWEN *Virg. Æneid* II. 236 Hempen cords cast over its neck.

fig. **1675** COTTON *Poet. Wks.* (1765) 297 Coarse hempen Trash is sooner read Than Poems of a finer Thread.

b. In humorous phrases and locutions, referring to the hangman's halter.

a **1420** HOCCLEVE *De Reg. Princ.* 454 Ware hem of hempyn lane! For stelthe is meeded with a chokelewe bane. *a* **1529** SKELTON *Agst. Garnesche* 162 Stop a tyd, and be welle ware Ye be nat cawte in an hempen snare. **1593** SHAKS. *2 Hen. VI*, IV. vii. 95 Ye shall haue a hempen Caudle then, and the help of hatchet. **1594** NASHE *Unfort. Trav.* 67, I..scapde dauncing in a hempen circle. **1606** DEKKER *Sev. Sinnes* VII. (Arb.) 44 Lamentable hempen Tragedies acted at Tiburne. **1632** RANDOLPH *Jealous Lovers* (N.), Shall not we be suspected for the murder, And choke with a hempen squincey? *a* **1700** B. E. *Dict. Cant. Crew*, *Hempen-widdow*, one whose Husband was Hang'd. **1785** GROSE *Dict. Vulg. T.* s.v., A man who was hanged is said to have died of a hempen fever. **1837** SIR F. B. HEAD *Narrative* viii. (1839) 208 What could they be worth to him but a hempen neck-cloth?

2. Resembling hemp.

1651 J. F[REAKE] *Agrippa's Occ. Philos.* 100 It makes a Hempen colour. **1772–84** COOK *Voy.* IX. IV. iii. (R.), Made of the bark of a pine-tree beat into a hempen state.

B. *sb.* Hempen cloth.

1777 ROBERTSON *Hist. Amer.* (1783) I. 255 They found Balboa..wearing coarse hempen used only by the meanest peasants.

†'**hempenly**, *a.* *nonce-wd.* [f. prec. + -LY[1].] Relating to or connected with hemp.

1609 PAULE *Life Abp. Whitgift* 40 A choise broker for such souterly wares, and in regard of his hempenly trade, a fit person to cherish up Martins birds.

hemph, obs. var. HUMPH *int.*

hempie: see HEMPY.

hemping: see HEMPEN.

hempland. Land appropriated to the growth of hemp; a piece of land formerly so applied.

1526 *MS. Acc. St. John's Hosp., Canterb.*, Rec...for ferme of hempland iiij*d.* **1670** EACHARD *Cont. Clergy* 93 A couple of apple-trees, a brood of ducklings, a hempland, and as much pasture as is just able to summer a cow. **1735** *N. Riding Rec.* IX. 131 The other closes and parcells of ground called Hemplands. **1846** E. SPURDENS *E. Anglian Words* (E.D.S.), *Pightle*, the little man's little field: called in Suffolk a *hempland*, without respect to the produce.

†**hempling**, *a.* *Obs.* Also 6 -lynne. [f. HEMP: cf. *hemping* = HEMPEN.] Of hemp, hempen.

1492 *Churchw. Acc. Walberswick, Suffolk* (Nichols 1797) 190 Two hempnling toweles. **1594** in *Archæol.* XLVIII. 136 Item v hemplynne square clothes.

'**hemp-nettle.** *Herb.* A name for the genus *Galeopsis* (N.O. *Labiatæ*), and esp. the common species *G. Tetrahit*; cf. *nettle-hemp* in HEMP 5.

1801 WITHERING *Brit. Plants* (ed. 4). **1861** S. THOMSON *Wild Fl.* III. (ed. 4) 251 Another lipped flower is the..hemp-nettle. **1863** BARING-GOULD *Iceland* 242 In the grass grew the common hempnettle.

hempseed ('hɛmpsiːd). The seed of hemp.

a caudle of hempseed = 'hempen caudle' (HEMPEN 1 b). *c* **1325** *Gloss. W. de Biblesw.* in Wright *Voc.* 156 *Canoys*, hempseed. *c* **1532** DEWES *Introd. Fr.* in *Palsgr.* 915 Hempe sede, *canebuise*. **1588** *Marprel. Epist.* (Arb.) 17 He hath prooued you to haue deserued a cawdell of Hempseed. **1694** *Phil. Trans.* XVIII. 36 Of a grey colour, and a convex figure, like the half of an Hempseed. **1714** GAY *Sheph. Week* Thursday 31 This hempseed with my virgin hand I sow, Who shall my true-love be, the crop shall mow. **1838** T. THOMSON *Chem. Org. Bodies* 429 Oil of Hempseed is obtained by expression from the seeds of..hemp.

b. A gallows-bird.

1597 SHAKS. *2 Hen. IV*, II. i. 64 Do, do thou Rogue: Do thou Hempseed.

c. *attrib.*, as **hempseed bird**, a bird fed on hempseed; **hempseed calculus** (*Path.*), name given by Wollaston to some varieties of the mulberry-calculus.

1611 CORYAT *Crudities* 15 Many gold Finches, with other birds which are such as our hempseede birds in England. 1864-70 T. HOLMES & HULKE *Syst. Surg.* (1883) III. 237 The dumb-bell crystals often unite into a mass and form the nucleus of a concretion called the hemp-seed calculus. *Ibid.* 246 The small, smooth, globular 'hemp-seed calculus'.

'hempstretch. *nonce-wd.* A person hanged. Cf. *stretch-hemp*, HEMP 3.

a 1843 SOUTHEY *Comm.-pl. Bk.* I. 369 One of the men who were hanging.. asked him.. to cut the rope. He did so, and Hempstretch fell on his feet.

'hempstring. *lit.* String or cord made of hemp. Hence *transf.*, one who deserves the halter.

1566 GASCOIGNE *Supposes* IV. ii, If I come neere you, hempstring, I will teache you to sing sol fa. 1606 CHAPMAN *Mons. D'Olive* Plays 1873 I. 241 A perfect yong hempstring. *Van.* Peace, least he overheare you! 1885 HOWELLS *S. Lapham* I. i. 40 He cut the heavy hemp-string with his penknife.

† **'hemptery.** *Obs.* Also **hemptre, -teren.** [? for *hempery, hempry.*] Hempen fabric.

1570 *Bury Wills* (Camden) 156, I beqwethe to my dawghter Jone.. one payer of shetes of hempteren.. to my dawghter Anne.. one payer of sheets of hemptery.. to John Kanam my sonne.. one payer of shetes of hemptre.

hempton, obs. form of HEMPEN.

† **'hemp-tree.** *Obs.* An old name of the Chaste Tree, *Vitex Agnus-castus.*

1548 TURNER *Names of Herbes* G viij b, Vitex is.. a tree and hath leaues lyke Hemp.. Wherfore it may be called in englishe Hemp tree, or Chast-tree, or Agnus tree. 1597 GERARDE *Herbal* (1633) 1388. 1611 COTGR., *Amerine*, Agnus castus.. chast or hempe tree.

'hempweed.

† **1.** Some kind of sea-weed; ? = DULSE. *Obs.*

1620 MARKHAM *Farew. Husb.* iii. 28 You shall gather from the bottome of the Rocks (where the seydge of the Sea continually beateth) a certaine blacke weede, which they call Hemp-weede, hauing great broad leaues.

2. = HEMP-AGRIMONY, and other species of *Eupatorium.*

1796 WITHERING *Brit. Plants* (ed. 3) III. 707 *Eupatorium cannabinum*,.. Hemp Agrimony, Dutch Agrimony, Water Agrimony, Water Hemp, Common Hempweed. 1862 ANSTED *Channel Isl.* (1865) 177 The hemp-weed or hemp-agrimony, a common plant enough. 1886 *Syd. Soc. Lex.* s.v., Aromatic hempweed, *Eupatorium aromaticum.* Round-leaved Hempweed, *Eupatorium rotundifolium.*

hempy, hempie ('hɛmpɪ), *a.* and *sb.* [f. HEMP *sb.* + -Y.]

A. *adj.* **1.** Made of, like, or of the nature of hemp; hempen; having or producing hemp.

c 1440 *Promp. Parv.* 235/2 Hempyne, or hempy.., *canabeus.* 1572 J. JONES *Bathes Buckstone* 10 b, Such [euill ayre] as commeth of Hempy grounds, as in Holland. 1611 COTGR., *Chanureux*, Hempen, Hempie, of Hempe. *c* 1645 HOWELL *Lett.* II. 54 'Twixt the rind and the Tree there is a Cotton, or hempy kind of Moss, which they wear for their Clothing.

2. *Sc.* and *north.* Worthy of the hangman's halter; usually jocular, meaning merely Mischievous, giddy, often in scrapes.

1816 SCOTT *Old Mort.* xlii, I was a daft hempie lassie then, and little thought what was to come o't. 1825 BROCKETT, *Hempy*, mischievous—having the qualities likely to suffer by cat o' nine tails, or by the halter. Applied jocularly to giddy young people of both sexes. 1885 RUNCIMAN *Skippers & Sh.* 110 Noted as the most 'hempy' boy in the.. district.

B. *sb.* One who deserves to be hanged; one for whom hemp grows. Usually jocular: A mischievous giddy boy or girl.

1718 RAMSAY *Christ's Kirk* III. xviii, He had gather'd seven or aught Wild hempies stout and strang. 1818 SCOTT *Hrt. Midl.* l, 'Where did you get the book, ye little hempie?' said Mrs. Butler. 1864 J. HARDY in *Proc. Berw. Nat. Club* 181 This hempie of a bird has taken to colonising. 1893 CROCKETT *Stickit Minister* (1894) 259 She had been a big-boned 'hempie' at the Kirkland School.

hemton, obs. form of HEMPEN.

NB. "*. . . whence the sexes were popularly mistaken*" (hemp col. 1). If so, a deliberate mistake surely, since the sex were called after the purpose and person to which th were allotted by custom. Thus, the female plant, and seed, went to the carl or husbandman for ropes, twin coarse work, and next year's planting: whilst the m plant, with its finer fibres, went to the huswife f homespuns, linen, lace—and probably simples a caudles also.

Text: “archive” 3 oz.sq.yd. 100 gsm
Cover: “card” 6 oz.sq.yd 200gsm

Part III. is set in Garamond.

Sample paper swatches with information and prices may be obtained from the publishers. Please sent three first or second class stamps.

treefree® hemp content, long life paper is Cht.’s original name and recipe for an entirely woodless paper, made only from annual indigenous plant fibres—ideally the unbleached ‘hurds’ or wastes—of true hemp, corn straw or flax. Guaranteed by our registered trade watermark, No. 1498203, at the Patent Office in April 1992. Universally fit, as evidenced by a lesser fibre below, for all writing and printing purposes!

“***C***oarse ***h***empen ***t***rash is sooner read
Than poems of a finer thread.
COTTON. *Poet. Works.* 1675. (pub. post.)